In Search of Howse House 1810-1811

1st Hudson Bay Company Trading Post South of the 49th Paralles & West of the Rocky Mountains

By
Carl W. Haywood

Published by Rockman's Trading Post, Inc.
Thompson Falls, MT

First Edition 2013

Copyright © 2012 by Carl W. Haywood

All rights reserved. No portion of this book may be reproduced in any form, or by any electronic, mechanical or other means without the prior written permission of the publisher.

Printed in the United State of America
Library of Congress Control Number: 2013901984
Cover Design: Flag of the Hudson's Bay Company. An 1800s painting showing Hudson's Bay Co. employees trading with Native Americans. (Image: Hudson's Bay Co./ Fulton Archive)
Published by Rockman's Trading Post, Inc.

Cataloging in Publication Data

Haywood, Carl W., 1942-Present

Howse House – 1810-1811,
The first Hudson Bay Company Trading Post in the Pacific Northwest

Includes bibliographical references.
ISBN 978–0–9802279–2–5: Paperback
1. Dedication. 2. Acknowledgments. 3. Preface. 4. Introduction. 5. The stage is set. 6. Let the game begin. 7. The Search for Howse House. 8. Conclusion.

Rockman's Trading Post, Inc
PO Box 2444
Thompson Falls, MT 59873
(406) 827-7625
www.rockmanstradingpost.com

Dedication

In the summer of 2010, my friend and mentor, Mark White, an archaeologist with the U.S. Forest Service, was diagnosed with terminal pancreatic cancer. Mark, my wife Linda and I had spent many weekends in the field and hours of discussions while I was researching and writing my book, Sometimes Only Horses to Eat, published in 2008. Mark was probably the most knowledgeable person around when it came to the subject of the fur trade. He spent almost a quarter century scouring northwest Montana attempting to locate and map fur trade posts and trails. He was especially interested in David Thompson and Joseph Howse. A few months before his passing in January 2011, Mark asked us to try to finish the research he had started by locating the Howse House site. This paper is the fruit of that labor Mark. Until we meet again along the last trail, thanks for being my friend.

Table of Contents

Howse Post Assignment i
Picture of Joseph Howse iii
Acknowledgments v
Preface vii
Introduction ix
    The Stage was Set 1
    Let the Game Begin 7
    The Search for Howse House 13
    Emma Ingalls 17
    Sam Johns 19
    Braunberger and White 23
    Ashley Creek Locale 29
    Peninsula Locale 35
    The Route 41
    One Final Mystery 43
Conclusion 45
References 49

Appendix 51
    A copy of:
    Howse House, An Historical and Archaeological Study
     – Barry Braunberger and Thain White, 1962

NAME: HOWSE (HOWES), Joseph   **PARISH:** Cirencester, Gloucestershire   **ENTERED SERVICE:** 22 May 1795 (A.1/47, fo. 57; A.32/17, fo. 48)   **DATES:** b. 13 February 1774[†]   d. 4 September 1852

Appointments & Service Outfit Year*	Position	Post	District	HBCA Reference
1795, 5 June-31 Aug.	travelled on *King George* from Gravesend to York Factory			C.1/398, 399
1795 - 1797	Writer	York Factory		A.30/7, fo. 26, 69
1797 - 1798	Writer	Gordon House		B.81/a/3
1798 - 1799	Writer	York Factory		B.239/a/101
1799 - 1801	Writer in charge	Carlton House		B.49/a/30; B.239/b/66, fo. 89
1801 - 1803	Writer		York	A.16/34, fo. 92
1803 - 1804	Writer	[Carlton House]		B.239/d/128
1804 - 1805	Writer	Chesterfield House		B.239/d/129
1805 - 1807	Trader	Carlton House		A.16/34, fo. 168; B.239/d/132
1807, Summer	Trader	Paint Creek House		B.60/a/6; B.239/b/74, pp. 38-40 Search File: "Howse, Joseph"
1807 - 1809	Trader	Carlton House		B.239/d/147, p. 128
1809 - 1810	Trader	Edmonton House		B.60/d/2B; B.239/d/147, pp.113-114
1810 - 1811	Master and Trader	first HBC Officer to cross Rocky Mountains		B.60/a/10
1811 - 1812	Inland Master	Paint Creek House		C.1/424, 425
1812, 13 September[○]	on the *King George* to London			C.1/778, 779, 780
1813	on the *Prince of Wales* to Churchill and York Factory			
1813 - 1814	Trader		Saskatchewan	B.239/a/124
1814 - 1815	Chief	Ile-a-la-Crosse	Saskatchewan	B.42/a/140; B.89/a/3
1815, 19 Sept.-5 Nov.	returned to England by *Prince of Wales*			C.1/785

An Outfit year ran from 1 June to 31 May

[†] In one of Peter Fidler's books (RB FTL PF 27) an entry reads "Joseph Howse natus fuit Feb. 13, 1774"
[○] While in London, Joseph Howse dealt with the legal papers surrounding the death of a relative (A.36/7, fos. 180-189)

Biography - *Dictionary of Canadian Biography*, Vol.VIII
Search File: "Howse, Joseph"; Genealogical records gathered by Walter K. Miles (E.376)

Parents: Thomas and Ann (DCB)
Wife: Mary, "an Indian woman" (Cree) (E.235/116; E.376)
Children: *son born 20 December 1798 (B.239/a/101)
Henry (b. ca. 1797-1808) (Search File: "Howse, Joseph" – descendants; E.376/1; E.235/116); m. Janet Spence 1 Nov. 1830 (E.376)
Jenny (Jane), b. 1812, baptised 29 April 1824, #352. Killed by a threshing mill of Mr. Bird's where she resided, buried 27 Feb. 1830 (E.4/1a; B.235/a/13; E.4/1b)

* possibly the same person

Filename: Howse, Joseph (1774-1852) (fl. 1795-1815). January 1988 JHB wg; rev. 2003/01

Joseph Howse post assignments from HBC records.

i

Joseph Howse
1774 –1852

Joseph Howse – circa 1844 – when he was 70 years old. From his book "A grammar of the Cree language, with which is combined an analysis of the Chippeway dialect" (London, 1844; repr. 1865)

Acknowledgments

When Mark requested that my wife Linda, Denny Rey and me continue his research with locating the site where Howse House once stood, he also provided us with information on the subject that he had collected over the years. Thanks too to Ron Beard for his assistance and advice. Without their interest and assistance this book would not be the same.

The most extensive research ever published on the subject appears to be a manuscript researched and written by Barry Braunberger and Thain White published in 1962, two years after I graduated from high school. At the time, I had little interest in the fur trade nor had I heard much about the subject except for passing comments made to me by Louis Caywood, father of my friends Rich and Rob Caywood.

Louis was the archaeologist who was then in charge of what is now the Besh-Ba-Gowah Archaeological Park located in Globe, AZ. In conversations with him I learned he had been responsible for the archaeological digs at both Fort Vancouver and Spokane House. Both were part of the North West Company's (NWCo) empire of fur trading posts during the 18th and 19th centuries.

It was during my research for my first book about fur-trader-explorer-surveyor David Thompson, Sometimes Only Horses to Eat, when I first stumbled across Caywood's reports.-They were extremely useful in my effort. Little did I know at the time how intense my passion for history would later become as I explored the vast influence and exciting conflict of the rivalry between the Hudson's Bay Company (HBC) and the NWCo as they struggled to dominate and control the trade in beaver pelts. Peltry was primarily used for making "Beaver Hats" in Europe. Their wide-scale battle continued well into the 19th century.

This manuscript is intended, as Mark requested, to collect as many available references as possible related to the location of Joseph Howse' House and to take our best shot at establishing its whereabouts. The question has remained controversial for at least 100 years.

We did not discover much new, earth-shaking information. Instead we gathered available information then applied our methodology and interpretation of where we think it all leads.

A special thanks to Linda for her efforts at editing my, sometimes awful, writing and above all for her patience with me.

I am not an archaeologist, or an historian, not even a very good writer but I hope you, the reader will in spite of this, enjoy my humble attempt at compiling and interpreting information on the subject. My interest in this effort is to search for the truth, wherever that may lead—nothing more.

Carl Haywood
Thompson Falls, MT.

Preface

In his introduction to Pinkerton's book published in 1931, Stewart Edward White defined well the gradual intrusion of what we now call "urban legend" into history.

> *"Ordinary history conveys information; extraordinary history conveys understanding.*
>
> *Unfortunately the capacity to understand is none too common. Its place is too often taken by personal prejudice, for or against. The result then is what we might call an historical legend. We are not particular enough in our examination of sources to distinguish between those that are genuinely such, and those that but repeat an error merely because it is to be found in early print. Thus one quotes from the other in long succession, each repetition lending to its strength, until at last the legend has taken the solidarity of indisputable fact."*

We believe that indisputable historical evidence is the only way to ultimately prove one's theories or suppositions about questions like the location of Howse House. This paper is our best assessment of where the most likely location for the post would have been based on available information. After examining the historical evidence, and based on our research and field trips to the area, we have concluded the precise location is not likely to be found due to factors explained and discussed later in this paper. However, we do believe that our conclusions are at least as good or perhaps even a little better than earlier attempts at solving the puzzle. Why will be explained.

For almost 100 years, debate has continued over the location of Howse House, a fur trade post erected and operated by Joseph Howse of the Hudson's Bay Company (HBC). Traditional sources tend to lean towards a location on Ashley Creek in the vicinity of present day Kalispell, MT. Others have suggested it was actually located on Lake Pend Oreille (pronounced Pond-Oh-Ray) near where Pack River empties into the lake between the present day towns of Sandpoint and Hope, ID. As I look at the evidence, I am convinced those who support the former are correct. If that is true, then the question becomes where on Ashley Creek was the post built?

Most surviving information dealing with the puzzle has likely been located. In 1962, long before public access to the Internet for research was likely even conceived, Thain White and Barry Braunberger published a comprehensive paper titled Howse House: An Historic and Archaeological Study. The paper includes references and quotes from most of the known journal entries and letters pertaining to Joseph Howse' 1810 foray across the Rockies and to clues they provide about the location of his trading post. It was of tremendous value to us during our attempt to determine where the post was most likely located. A copy of their paper dealing with historical information and their conclusions can be found in Appendix A.

Introduction

This manuscript is primarily about Joseph Howse, the first HBC fur-trader to cross the Rocky Mountains in search of the wily beaver and the builder of the first HBC post on the west side. At the same time it is also about David Thompson, NWCo fur-trader, explorer and surveyor. The two rivals represented the two largest fur-trading companies in the New World. Both men were experienced traders. Both had worked their way up through the ranks to become trusted company men. They were almost like two peas in a pod. Each was fiercely loyal, highly experienced and very competitive. Each knew his way around the unexplored western wilderness where furs were more valuable than money. In the eventful year of 1810, Joseph Howse was 30 years old—David Thompson was 42.

Joseph Howse was a fur trader and linguistic scholar. Baptized March 2, 1774, in Cirencester, England, he was the son of Thomas Howse and his wife Ann. After ten years in the fur trading with the HBC in the Saskatchewan district, Howse took part in exploration of western North America. He was the first Hudson's Bay Company man to cross the Rocky Mountains; a pass, a peak and a river bear his name. Howse is also remembered for his epic grammatical work that resulted in the translation of the Cree language into English.

The first stage of Howse' apprenticeship ended in December 1797, when as a writer and accountant he left York Factory for Gordon House on the Hayes River. He spent the rest of the season there and did not return to York Factory until July 4, 1798. David Thompson obviously knew Howse since he gave Howse a letter to be delivered to James Hughes at Fort Augustus. Doing so is a strong suggestion that Company Men intended no exploration farther west than the foot of the Rocky Mountains at the time.

As the HBC prepared to challenge the position of the NWCo in the Rocky Mountains, Howse joined James Bird at Edmonton House (located at present day Edmonton) in 1809, and with the exception of his journey across the mountains he remained at that assignment until 1811. Howse was back at Edmonton House by mid July 1811, the first HBC man to have followed the NWCo into the land across the Rockies. According to HBC records, following his return in 1811, he was assigned to the position of "Inland Master" at Paint Creek House. He remained there only a short period of time. HBC records note that on September 13, 1811, he was aboard the ship *King George* on his way to London.

Like David Thompson, Howse was a mapmaker. HBC journals, letters or other documents credit him with three drafted maps and he collaborated on a fourth with William Auld at York Factory. Auld extolled Howse for his bravery, leadership and success as a trader for which he was rewarded with a gratuity of £150 from HBC leaders.

David Thompson was born April 30, 1770, near Westminster, England and educated at the Grey Coat Mathematical School for underprivileged boys where he was introduced to basic navigational skills. They would serve him well in the future. In 1784, when he was only fourteen years old, he was engaged—*indentured* would be a better description of the relationship—by the Hudson's Bay Company (HBC). He worked for the HBC from 1784 until 1797. Unhappy at not being allowed to explore and map new country he abruptly resigned and immediately signed on with the Montreal-based NWCo where he eventually became a partner.

In 1807, the NWCo sent Thompson across what came to be known as Howse Pass. This was the first time he crossed the Rocky Mountains. In 1808, he built Kootenai House near the mouth of Toby Creek a few miles north of Athalmer, British

Columbia. That same year, Thompson made an attempt to reach the Saleesh River (today's Clark's Fork) making it as far as the mouth of Deep Creek, a few miles west of Bonners Ferry, ID. His plan was to follow a major Indian road leading in a southerly direction until he reached the Saleesh River. He referred to the road by several names including the "Great Road of the Flat Heads" and "The Lake Indian Road".

Thompson quite obviously had a guide who was very familiar with Indian travel routes in the region. Facing several obstacles that ultimately foiled his plan to reach the Saleesh River that year he decided to return to Kootenai House for the winter. The guide, whoever he was, guided Thompson back to the north by way of a northerly route up the Moyie River. From the headwaters of the Moyie they crossed a divide—a 'height of land'—to get back to the Columbia River basin. The route would later become one of the main trails from the south used by miners to reach newly discovered goldfields of British Columbia in the vicinity of Fort Steele.

In 1809, Thompson established two new trading posts in what is now northern Idaho and northwestern Montana. Both were located on the Saleesh River. The former, Kullyspel House, was located on a peninsula that jutted out into Lake Pend Oreille near the present town of Hope, ID; the latter, Saleesh House, a few miles east of Thompson Falls, MT.

A year after Thompson established his two posts on the Saleesh River, the HBC decided it was time to have a look at the potential of the new beaver hunting grounds. As was usual, when one of the companies moved to a new area, it was not long before the other moved there too. In fact, rival posts were often within hailing distance from one another. In a few cases, posts actually shared palisade walls to provide common protection from mutual enemies.

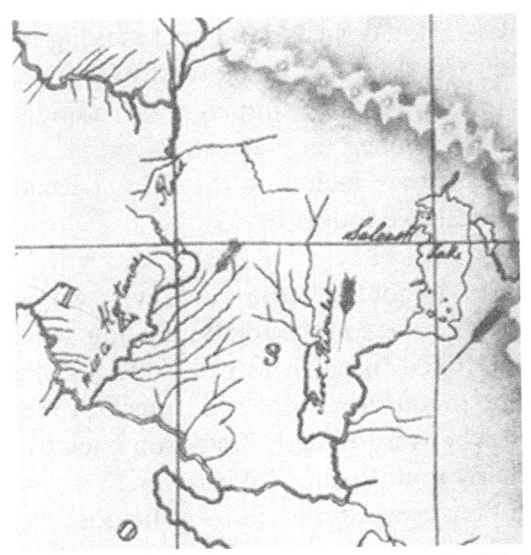

Segment of David Thompson's 1814 Map of Saleesh Lake.

HBC's plan was to send Joseph Howse across the mountains to locate Thompson and "spy" on him. His goal was to construct a small trading post there, spend the winter of 1810-1811, exploring the region and engage in fur trade with the indigenous people in the area. Based on his experience through the winter, he would provide the HBC with his opinion as to whether or not the company should continue a long-term trading house in the region.

Howse crossed the Canadian Rockies via a pass later named after him. To achieve that goal, he needed to locate Thompson's posts and see what Thompson was up to and how well he was doing. His venture was the first time the HBC had been in what Simon Fraser called *New Caledonia*, which included most or all of the Province of British Columbia. The northwest region of what is now the United States was referred to as the *Columbia District* and was where David Thompson built two of his trading posts.

In his book titled *A History of the Canadian West, 1870-1871*, Arthur Silver Morton wrote that:

> "*Before the merger* [with the NWCo in 1821] *the Hudson's Bay Company had done little but explore the possibilities of trade in the Far West. Representing the firms during 1810-1811, Joseph Howse had, however, crossed the Rocky Mountains at a pass now bearing his name. He had done so at the risk of being killed either at the hands of the warring Piegan Indians or by rival Nor'westers. Howse had ventured as far west as the Flathead country and he returned with glowing accounts of his experiences in which the dangers of his journey were not emphasized. Why should they have been? He brought back thirty-six bundles of furs on which he gleaned a profit of 75 percent. But with the exception of this isolated exploit, the new Hudson's Bay Company was obliged to draw upon the experiences of those who had represented the North West Company in the Pacific Northwest.*"

Morton was mostly right – with the exception of the part about Howse returning to Canada,

> "*...with glowing accounts of his experiences in which the dangers of his journey were not emphasized.*"

The rewards looked good. Trading goods, stores and wages for Howse' expedition had cost the HBC £576. The furs brought back were valued at £1,500. In today's dollars, and adjusted for the current exchange rate and inflation since 1810, the cost of the expedition was about $5,056 and the furs valued at $13,175.

That's a return of more than double the initial investment, not bad for a few months' work.

However, it was the part about *"the dangers"* of conflict with the Blackfeet, more specifically the Piegans, which caught the attention of the HBC leadership. This led to a decision not to return to the west side of the mountains to pursue the fur trade. The HBC had been trading partners with the Blackfeet for decades. At the time, they did not deem it wise to jeopardize that relationship.

The HBC and NWCo were intensely competitive adversaries in the fur trade. By means of their trading houses east of the Rocky Mountains *both* companies traded with the tribes of the Blackfeet Confederation for decades. During that long relationship the Blackfeet obtained guns that provided them the means to dominate tribes west of the mountains if they wished. Among these tribes were the Salish (Thompson's Saleesh), Kootenai, Pend d'Oreille, Spokane (Thompson's Spokan) and Kalispell (Thompson's Kullyspel)—tribes still reliant on stone-age weapons technology including obsidian blades and points for arrows, spears and knives; weapons used for hunting as well as for defense from their enemies. Trade with HBC and NWCo would undoubtedly introduce guns and iron points to these traditional enemies of the Blackfeet. Retaining that power meant stopping traders from crossing the Rockies.

In the spring of 1810, Thompson was bound for Rainy Lake House with furs and pemmican produced during the previous winter. The NWCo's Rainy Lake supply depot was located on the Canadian side of the International Boundary a little west of what is known today as International Falls, MN.

June 23, on his way down the North Saskatchewan Thompson passed Howse' two canoes bound for the Columbia. Howse' group had started up the river from Fort Edmonton on the 19th. Thompson noted that the encounter had taken place *"...5 leagues below Old Fort Augustas"*. He was unaware at the time that his competitor had orders to locate his posts on the west side of the mountains and see what he was up to. If Thompson had discovered a good fur producing area, the HBC wanted in on the action.

The stage was set.

Drawing by Cricket Johnston, 2007

The Stage was Set

The itinerary of the HBC's first foray across the Rocky Mountains is, ironically, documented only in the journals of David Thompson and Alexander Henry of the NWCo. Unfortunately, if Howse kept a journal, which he likely did, it has been lost. As a result we have but scant information about his travels and where he built the post bearing his name.

Starting from their neighboring posts of Edmonton House and Fort Augustus, the two explorers set off on July 18, 1809. The HBC party consisted of Howse and three others, and its purpose was obvious. In Thompson's words, Howse *"went off for the Mountains to examine the Country &c&c."* When Thompson and his party, preparing to winter in the mountains and thus encumbered with supplies and trading goods, reached Rocky Mountain House on July 26, Howse and his men were already several days ahead. The two parties did not meet until August 9, at the head of navigation, at the forks of the North Saskatchewan River *"...when Mr Howse & the Indian with him..."* were already on their way back from the pass.

During the winter of 1809–1810, while Thompson was trading on the Saleesh River, James Hughes and Alexander Henry of the NWCo, decided it was in the Company's best interest to consolidate the upper "forts des prairies", Fort Augustus II and Fort Vermilion. To replace them, a new post would be constructed at "Terre Blanche" (White Earth), near the confluence of the North Saskatchewan River and White Earth Creek, about half way between two previous posts. The new post was located west of the confluence of the North Saskatchewan River and White Earth Creek, near Pakan, Alberta. It was to be shared with the HBC. The result ended in an interesting arrangement between the rivals—two separate, independent trading houses that were surrounded by a common palisade. Inside, there was a fence dividing the enclosed site into two distinct areas. The NWCo's White Mud House was constructed on one side of the fence, the HBC's Edmonton House on the other. The purpose of combining locations was in the interest of a stronger mutual defense against attacks by hostiles that neither would have been able to muster on their own:

> *"The North West Company fort, specifically described in the journals of Alexander Henry, contained an "Indian Hall" and storehouse of two storeys [70 x 20 feet], a blacksmith's shop, the two-storey Master's house, houses for the men and outbuildings. The Hudson's Bay Company portion of the fort, which was separated from the North West Company portion by a fence across the enclosure, probably had similar types of structures. Bastions lay at the southwest and northeast corners of the palisade."*

Since the rivals had, by then, combined their posts within a common palisade, Henry was in a position to observe the activities of the HBC. On June 19, 1810, he recorded the departure of Howse' brigade from the adjacent post thusly:

> *"...two canoes for the Columbia, with nine men... They embarked four rolls of tobacco, two kegs of high wine, powder, several bags of balls, a bag of shot, pemmican, etc."*

So it was that Howse began his journey from Edmonton House to the Columbia River. As Henry noted in his journal on June 20, Howse had departed:

> "...with four Cree guides and hunters . . . the whole HBCo. Columbia expedition consists of 17 persons."

The Howse brigade had two goals—to trade for beaver pelts and to provide time for Howse to do some exploration. The NWCo took the challenge seriously and on July 9, James McMillan *"set off for the Columbia to watch the motions of the H.B. in that quarter"*.

Most credible historians believe the only record of the journey by Howse' own hand was contained in a letter written to HBC governor Sir George Simpson in 1843. In sketching his itinerary Howse wrote that he:

> "...*crossed the Rocky Mounts in the Summer and Autumn of 1810 by ye North Branch of the Saskatchewan – ascended the Kootoonay* [Columbia] *River – carried into the Flat-Bow* [now the Kootenai] *River – descended by the most Southly Bight of it – crossed* [Portage Poil de Custer] *to Flathead River . . . where we built."*

This 1859 map shows the location of Howse House to be north of Flathead (Saleesh) Lake. Keep in mind that this map was produced nearly 50 years after the post was constructed near the junction of two major Indian roads. One led across the divide to the west then down Wolf Creek to its confluence with Fisher River. The other followed the Flathead River to the north where it crossed the divide between the Flathead and Kootenai Rivers then followed along Dunn Creek to the Kootenai (McGillivray's) River.

Howse stated that he had crossed the *"Portage Poil de Custer"* (sic) located a short distance above the Big Bend. Thompson's

Poil de Castor portage was likely located in the vicinity of the point where Dunn Creek empties into the Kootenay River. (Note: Names in italics were added for informational purposes.)

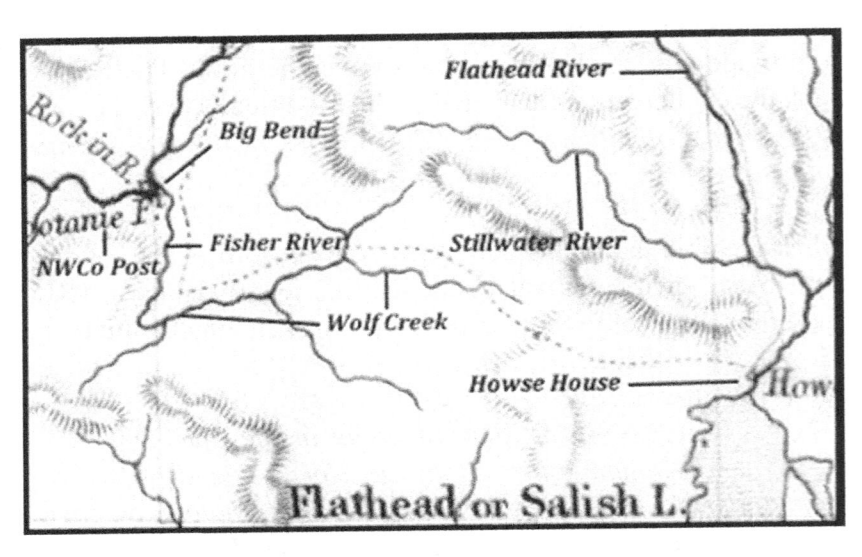

This 1859 map shows the trail from Howse House on Flathead River to the divide near the head waters of Wolf Creek to its confluence with Fisher River then down Fisher River to the Big Bend.

Mark and I had several discussions about the travel route both on the ground and in follow-up conversations. We were in general agreement that it was highly likely that Howse had been misled by his guide and ended up in the Flathead Valley instead of on the Saleesh River where Thompson's posts were located.

A main Indian road from Big Bend led to the south up Fisher River, crossed Wolf Creek, turned up Snell Creek then fell back onto Fisher River a short distance northeast of Upper Thompson Lake on State Highway 2. That location is the "height of land" between the Fisher River and Thompson River drainages. More detailed information about the probable location of the road can be found in *Sometimes Only Horses to Eat*.

At Thompson Lakes there was a major fork in the road. One would have led Howse to the Saleesh River via the Thompson River, referred to as the "House Rivulet" by Thompson, where Saleesh House was located. The other fork led to the east generally following the Highway 2 route down Ashley Creek into the Flathead Valley. However, one must be cautious in using old maps. During my research I have found several maps drawn during the 1850–1860 period that include place names and trail locations that are clearly incorrect.

Howse' reference to crossing "Portage Poil de Custer" suggests he may actually have struck out up the Dunn Creek trail in very close proximity to the location described by Thompson as the "Poil de Castor". The route up Dunn Creek appears on several GLO and USGS maps. The trail follows Dunn Creek to cross the divide at Sugarloaf Peak then drops into Wolf Creek very near the old Indian Road crossing which Thompson used in 1809. The Dunn Creek route conflicts with the 1859 map.

The 1859 mapmaker was, in some way, familiar with the spot where the post had been located. The information used in drawing the map was quite possibly based on earlier maps, and could have erred by simply *assuming* that Howse' route from McGillivray's River was along Wolf Creek. There is little historical evidence to verify Howse' route to the Flathead River so the best we can do is attempt to follow the route he described to arrive in the Flathead Valley.

During his winter on Flathead Lake, Howse described a crossing of the continental divide in December 1810, probably to the Three Forks area. He wrote:

> *"...with a couple of my men I accompanied the Flat-heads to the head branches of ye Missouri – returned to our House – in Feby 1811."*

As we know, the *headwaters of the Missouri* are where the Madison, Gallatin and Jefferson Rivers converge in the vicinity of the town of Three Forks, MT. Earlier a small, remote trading post had been established there by American fur trader Manuel Lisa. It was ultimately abandoned following the Blackfeet's continual attacks that resulted in the deaths of several trappers. One of the well-known events associated with the post was the escape of John Colter from certain death at the hands of the Blackfeet. Colter was allowed to live only on condition he, given a short lead, was able to escape pursuing captors by literally running for his life.

Let the Game Begin

The departure of both the Howse brigade and Thompson brigade from the Edmonton House/Fort Augustus Post marked the beginning of the cat and mouse game that was to take place between the rivals.

Information dealing with the controversy about the location of Howse House is available in libraries and archives in the United States, Canada and England. The HBC's service record for "Howse, (Howes), Joseph", shows him as the "Master and Trader… first HBC Officer to cross Rocky Mountains" in 1810-1811.

Howse was dispatched on his mission for the HBC for the purpose of determining whether or not the west side of the mountains would be a good place for a fur trading post. Howse departure from Edmonton House did not escape the notice of the NWCo. Thus, James McMillan and Nicholas Montour of the NWCo left Terre Blance House on July 9, 1810, to follow Howse and keep an eye on his whereabouts. James Bird wrote:

> "*Received Letters from Mr. Howes, dated Cootana* [Columbia] *River 20th Augut. 1810, in which he says that, having been Informed by some Cootanaha's that a Battle had been fought between a party of Flatt Head Indians, with whom a Mr. Macdonald* [Finan McDonald] *clerk to the N.Wt. Coy, was in company, and a party of Muddy River Indians, in which the later were defeated with the loss of 14 Men killed; and that the Muddy River Indians in Consequence were laying in Ambush to intercept him, or any white Man, who might attempt to convey Goods to the*

Flatt Heads, he had determined on remaining some time at the place where this Letter was dated to gain further intelligence, after which, he should determine on his future proceedings. The News Mr. Howes had received has been seriously confirmed by the Muddy River Indians themselves who stopped 4 French canoes, which were bound for the Columbia, a little above old Acton House; and this band of Indians that stopped the Canadians declare that, another is Laying on the banks of the river [called by Mr. Thomson Macgillveray's River] *which Mr. Howes intended to decend, to intercept them there in case they had passed Acton House without their knowledge. I am consequently not with serious apprehentions for the Safety of Mr. Howes & party without a probability of being able to render them any assistance or, having it in my power to obtain any intelligence of their present situation. Mr. Thomson it appears after a Settlement had formed & Goods left for the Muddy River Indians at Acton House set off privately with twenty men & Horses conveying a few Goods, to go by another way towards his last years abode near the Columbia...."*

In his March 13, 1811, journal entry, James Bird provided additional information about the expedition based on a report provided by Howse himself:

"I hasten to answer your queries as follows:

"...I crossed the Rocky Mounts. In the Summer & Autumn of 1810 by the North Branch of the Saskatchewan, Ascended the Kootoonay

[Columbia] *River-and carried into the Flat Bow* [Kootenai] *River-descended the most Southly Bight of it* [the Big Bend of the Kootenai River] *crossed* [portage poil de Custer] *to Flathead River* [above the Lake-see Arrowsmith "Howses Ho."] *where we built. In Decr. with a couple of my Men I accompanied the Flat-heads to the head branches of the Missouri — returned to our House in Feby. 1811 and to the Saskatchewan the following summer."*

In the same entry, Bird provided information about important (and apparently successful) Company efforts to support of the expedition in the face of the Piegan blockade of the route across the mountains. As he put it, the strategy was based on:

"...*A desire to conciliate the minds of the Indians, and to dispose them as much as in our power to behave friendly towards Mr. Howse, and party should they, as is too probable, meet with him in their war Excursions, has induced us, and the Canadians, to comply with the urgent Solicitations of the Indians to leave Men and Goods at this Settlement during the Summer; and on this Acct.* [and a few freemen making their Spring hunts, in that Quarter] *only could this place have been thought worth maintaining. The Muddy River Indian Chiefs have Promised not to molest Mr. Howse on his return from the flat head Country; but declared that, if they with a white man going to supply their Enemies, they would not only plunder & kill him, but that they would make dry Meat of his body...*"

Evidently the Piegans had still not forgotten the defeat of their war party by west-side buffalo hunters accompanied by Thompson's men the previous summer. Their reaction and determination to keep NWCo., and apparently HBC traders as well, from crossing the Rockies continued to be problematic.

Thompson arrived back at White Earth House from Rainy Lake House by canoe on September 6, 1810, almost three month after Howse' departure. At 8:00 AM the following morning, his other three canoes arrived. Each carried six men and seventeen, 90 pound "pieces" of supplies and trade goods bound for the Saleesh River. The members of his brigade with the assistance of Henry's people spent the rest of the day making repairs to the canoes. At 4:00 PM on September 8, they were back on the Saskatchewan. Thompson sent his canoes on ahead since he planned to continue his travel upriver by horseback from White Earth House. He departed on September 11, three days behind his canoes. More information about White Earth House can be found on the website of *Canada's Historic Places*.

It was during his stay at Henry's post that Thompson learned that while he had been absent from the Saleesh River posts, three of his men, Finan McDonald, Michel Bourdeaux and Babteste Buche had accompanied the Flatheads into Piegan territory to hunt buffalo. During his first trading season at Saleesh House during the winter of 1809–1810, Thompson traded 20 guns to the west-side tribes. He and his brigade also took the time to teach them how to use the firearms effectively.

For many years, the Blackfeet had been sending war parties across the mountains in attempts to drive the tribes there farther west or, better yet, to exterminate them. For decades they had been armed with guns traded to them on the east side of the Rockies by the French and Dutch as well as by the HBC and the NWCo. The firepower of the Blackfeet enabled them to dominate tribes to the west with impunity since their arsenal at the time still consisted of spears, bows and arrows tipped stone points.

When the Blackfeet recognized Thompson's men among the Indian buffalo hunters they became very angry. Their old enemies were now armed. They attacked. According to Mark White:

> *"Finan McDonald and the Salish armed with 13 trade guns inflicted a defeat that the Blackfeet Indians were not going to forgive."*

The outcome of the confrontation was the defeat of the Blackfeet by the west-side tribes for the very first time. A total of sixteen Blackfeet were killed during the fight. An interesting side note here is that apparently Thompson's men were not the crack shots romantic history would sometimes have us believe about "mountain men." In Thompson's narrative he was brutally honest when he wrote:

> *"...Mr. Finan McDonald fired forty-five shots, killed two men and wounded one, the other two men fired 43 balls, and each wounded one man."*
> (See, *Narrative*, Page 425.)

The repercussions from that defeat were a determined effort by the Piegans to stop NWCo traders from crossing the mountains via Howse Pass.

After receiving, and accepting, a strong, stern warning by the Piegans about trading guns to their enemies to the west, Howse was allowed to continue his journey. Thompson's brigade, traveling several days ahead of their leader, with whom they had no communication at the time, were left with no choice but to find a way to circumvent the blockade. Finally, following several attempts to get around the blockade, and with a lot of help from fellow Nor'Wester Alexander Henry, they were finally able to sneak past the Piegans.

Thompson, with only a handful of men, was too late. He was forced to find a new route across the mountains to the Columbia River. Thus began his overland journey from the Saskatchewan River to the upper Athabaska. There he encountered intensely cold, winter weather and great challenges including starvation and mutiny by some of his men. He did not get back to Saleesh House until the summer of 1811. By that time, Joseph Howse and the HBC brigade were gone. The HBC did not establish another trading post west of the Rockies until 1821.

The Search for Howse House

Historic records and maps indicate that Howse House was probably located a short distance southwest of present day Kalispell, MT. In an earlier manuscript, Mark White noted that:

> "...three Arrowsmith Maps from three different decades show three different locations. The earliest 1811 Arrowsmith Map shows Howse House north of Flathead Lake and north of the Whitefish/Stillwater and Flathead River confluence. This location corresponds with Township 28 North, Range 21 West, sections 3 or 4..."

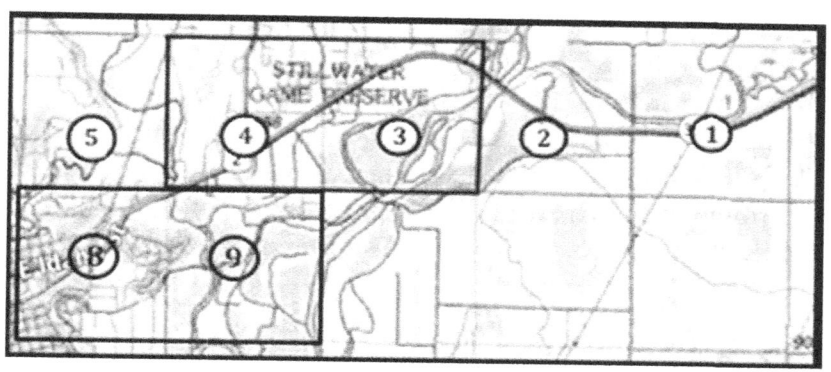

USGS quad, July 1, 1976 – Sections 3&4, T28N R21W

Note that much of this area lies in the floodplain of the rivers. Not a likely spot for a trading post made of timbers and boards. More details about this location as the possible site for Howse House is discussed in the following subsection titled *Braunberger and White*.

Mark went on to say:

> "The 1824 Arrowsmith Map shows the location of 'Howe's Ho.' below and west of the confluence of the three rivers in sections 8 or 9. The Arrowsmith Map of 1859, compiled from original documents from J. Arrowsmith, shows the location of Howe's Ho. south of the confluence of the previously named rivers and closer to Flathead Lake at the confluence of two early prehistoric/historic trails."

During our field visits we looked at all three potential sites. We continued the discussion for some time and pretty much concluded that due to the location and character of the terrain in sections 8 and 9 it was not a likely spot for a post other than a very temporary one.

We will probably never know for certain but it is important to note that the 1811 map would have been completed within a very short period of time after Howse return from the west. Or, he might have dispatched a map to the HBC soon after he determined where he would build the post but before he became familiar with the geography of the surrounding area.

Determining the precise location of Joseph Howse' trading post is difficult to say the least. Unlike Thompson, Howse was not provided with navigation and surveying instruments. In spite of that, he still drew several maps that were sent to the Governor of the Hudson Bay Company. These maps were:

> Map #58 "Rocky Mountains Sketch of shewing the connection of the Athapescow, Saskatchewan & Missouri Rivers by J. Howes 1812 and No. 82 "Columbia River Sketch of by Josh. Howes 1815."

Unfortunately those original field maps were evidently sent to the Arrowsmiths, the map-makers from London, and never returned but were referenced as existing in the Hudson Bay Company Catalogue of Archives. None of Howse' journals, as far as we know, have survived. What we do have is a letter written to Sir George Simpson in 1843, outlining the route Howse took in the summer of 1810. The reason this letter was written was Simpson was dealing with the Oregon Territory Boundary issue at the time. The letter included questions about Howse and McMillan's early crossings of the Rockies. On February 9, 1843, Howse wrote:

> *"Descended the most Southly Bight of it* [the Big Bend of the Kootenai River] *crossed portage poil de Custer to Flathead River [above the Lake; see Arrowsmith] where we built. In Decr. with a couple of my Men I accompanied the Flatheads to the head branches of the Missouri — returned to our House in Feby. 1811 and to the Saskatchewan the following summer."*

Descendents of Joseph Howse stand at the edge of what was likely "Mr. Griffith's grain field". View is to the northwest. L-R: Jennifer Howse, Kathy Howse, Lynn Roden, Wayne Howse Cree Hunter by Edgar Samuel Paxson, 1905

Cree Hunter by Edgar Samuel Paxson, 1905

Emma Ingalls

Emma and her husband, Clayton Ingalls, moved to Corvallis, Montana Territory, in 1886. In 1889, they moved to Demersville near present day Kalispell in what is now Flathead County. Emma and her husband founded the Kalispell Inter Lake Daily. By age 29, Emma had risen from reporter to editor to publisher of the newspaper.

The Ingalls' property was located near the Eugene McCarthy homestead. Joseph Ashley had originally claimed the site of the McCarthy parcel in 1857. In 1891, McCarthy ultimately patented the homestead. As a result, and based on Emma's experience as a reporter, she was no doubt familiar with the history of the area. In 1945, she wrote a paper titled *Over Trails of the Past*. In it she noted that:

> *"Ashley Creek was named for Joe Ashley who settled on the old Eugene McCarthy place south of Kalispell in 1857. He left here for the reservation in 1883 and Eugene McCarthy who had come in as a laborer on the N. P. Ry. then took the place. The remains of the old trading post established in about 1808 were still in evidence. Eugene McCarthy,* [Police Magistrate] *with the scrapers and mules his father had, salvaged from the construction work on the N. P. Ry. Tore down the old chimneys and filled in the excavation, and where this historic building once stood Mr. Griffith has a grain field."*

Emma's reference to *"Mr. Griffith's grain field"* was, without doubt, a reference to William Griffith who acquired the old McCarthy homestead in 1910. Her reference to the remains of an old trading post was probably Howse House although the date is obviously incorrect since Howse did not arrive in the region until 1810. As far as I have been able to find, this is probably the best clue as to the general location of the post that exists.

This 1814 map by cartographer Aaron Arrowsmith placed the location in the same vicinity described by Ingalls. Arrowsmith produced his first map of North America in 1795, based on maps and surveys provided by the Hudson's Bay Company from its archives in London. As far as we know, Arrowsmith was never west of the Rocky Mountains and thus was not personally familiar with the country. We do know that much of the information he used in his early maps was based on David Thompson's maps and notes.

1814 map by cartographer Aaron Arrowsmith placed Howse House in the same vicinity as that described by Ingalls.

Sam Johns

In 1943, Sam Johns wrote a paper titled *Early Settlers of the Upper Flathead Valley*. In it he wrote about the system of old Indian roads on the west side of Flathead Lake and in the lower valley. His description of an early HBC trading post existing on the lower west side of the valley strongly suggests the location may well have been on or very near the location of Howse House. If so, it would have been in the same vicinity as the old post described by Emma Ingalls. Johns wrote:

> "The Hudson Bay Trading Company established a trading post in Flathead Valley along about the year 1844 and placed as Manager therein Mr. Angus McDonald. This Post was located in the lower West side valley, near where the Brocken shhool [sic] house is now standing and was maintained for several years, after which it was transferred to Post Creek, on the Flathead Indian Reservation, and conducted by Mr. McDonald for many years. This change was made necessary by the fact that much of the trade with the Indians was being lost as they passed over the mountains to Eastern Montana trading posts via the Missoula River route."

Johns' account provides food for thought but appears to be historically incorrect in relation to Fort Connah. Construction of the post started in 1846, under the supervision of Neil McLean McArthur. In 1847, Angus McDonald was appointed chief trader at the post where he remained until its closure in 1871. McDonald's son Duncan served as its last factor.

Fort Connah (Peter Toffi)

Following the closure of Fort Connah, McDonald acquired the old post and surrounding lands. The McDonald family still owns the property and Angus is buried a short distance from the site. Fort Connah was the last HBC trading post in what is now the United States. The Fort Connah Preservation Society is in the process of restoring the old post.

The trading philosophy of the HBC and that of the NWCo were exact opposites. The HBC operation depended on beaver hunters bringing their pelts to a trading post to be exchanged for trade goods. In some places that system could mean days of travel. On the other hand the NWCo established posts where hunters in the area could bring their furs but also sent traders into the field to locate hunter's camps and make the trade there thus saving considerable time for the hunters.

During the Fort Connah period, as the fur trade began downsizing due to silk replacing beaver for top hats, smaller trading posts were established at sites close to hunting grounds. Over the years the remains of several of these old posts have been found.

I suspect the post referenced by Johns was likely one of those. It was most certainly not Fort Connah.

Recall that Thompson had sent James McMillan to find out where Howse had located his trading post. There can be little doubt that he found Howse and his men and reported back to Thompson about where the post was located. He probably also provided a description of what the post was like. Somewhere in some dusty attic or archives there may well be notes or letters about McMillan's report still waiting to be discovered. All we know for sure comes from Thompson's journal where he noted that McMillan reported that Howse was *"...in the Lake"*, undoubtedly a reference to Flathead Lake, called Saleesh Lake by Thompson in his journal and on his maps.

Pictographs like these were not uncommon along old Indian roads in the valleys of the Northwest traveled by early fur traders.

Braunberger and White (B/W)

During the past few decades modern technology has radically changed a researcher's ability to locate and access extensive information relatively quickly compared to that available a couple decades ago.

In 1962, Dr. Barry Braunberger and Thain White published a paper that is probably the most extensive and comprehensive document available on the subject of Howse House. It is titled Howse House, *An Historic and Archeological Study.* (See Appendix).

In the mid-1960's, Patrick Best—a descendent of Joseph Howse—working on a family genealogy project, contacted the Kalispell Chamber of Commerce regarding the post. He later wrote:

> *"Shortly thereafter, I received a letter from A.B. Braunberger, a practicing optometrist, in Kalispell, whose hobby was digging into the past. He told me that in the late 1950's, he had embarked upon a one-man crusade to learn more about the fur trading post Howse supposedly constructed in that area. After some years, he discovered he had "bitten off too big a chuck [sic] to chew". So, he enlisted the assistance of a self-educated anthropologist-archeologist-historian, Thain White, who ran a museum called "The Lookout on Flat Head Lake". In the course of their research, they learned that if they were going to get anywhere on this project, they would have to obtain the cooperation of the Hudson's Bay Company in London.*

> *"After some correspondence with the University of Montana and Washington State University, Mr. Braunberger wrote to Mrs. Alice M. Johnson, Hudson's Bay Company archivist. To make a long story short, he said before they were through, they had spent about $100.00 for research on Joseph Howse. It took him and Mr. White about four years in their spare time to compile the material on Howse and set it down in print."*

In the Preface of the B/W paper it states:

> *"In all, a total of eleven sites are listed under the archeological section, and these areas of investigation are presented because the topography, the convergence of Indian Roads, the early settlements, reports of artifacts, and information from informants made these sites likely prospects for investigation."*

There are two primary parts to the B/W investigation that appear to be *hypothetical* in my opinion. The first, that Howse House was located on Lake Pend Oreille requires a large leap based on pure speculation to make the connection. The second is that their investigation focused primarily on an assumption that Howse House was located on the Flathead valley floodplains.

Regarding the first hypothesis, a quote by Coues, incorporated in the B/W report, noted that on January 26, 1811, James McMillan and Nicholas Montour, another NWCo clerk, reached White Earth House:

> *"...from Flat Head Lake, which they left Dec 12th..."* and where *"the H.B.Co., were also settled."*

Assuming Coues was correct, it becomes clear that Howse post was not on or near Lake Pend Oreille, Thompson's Kullyspel Lake. The B/W paper includes the following:

> "*A map of the Columbia River by Alexander Ross dated Red River Settlement, 1 August, 1849, is in the collection of Additional Manuscripts in the British Museum...This map located, in spite of Ross's remarks quoted above, 'Mr. Hows 1810' on the west shore of Flathead Lake, towards the northern end...*"

Alexander Ross was with the Pacific Fur Company at Fort Astoria in 1811. He joined the NWCo in 1813, after they acquired the fort and re-named it Fort George. Following the amalgamation of the NWCo with the HBC in 1821, Governor George Simpson persuaded Ross to remain with the newly formed company (HBC) for a few more years. Ross led the 1824 Snake Country Brigade out of Flathead Fort to explore and trap the streams of western Montana and southeastern Idaho.

He left Flathead Fort for the Snake River Country in February 1824. He described his travels through *Prairie de Cheveaux* (Horse Plains, now Plains, MT), through *Prairie de Camas* (Camas Prairie). From there he proceeded in a southerly direction up the Bitterroot Valley and passed through what is now called Ross's Hole. Crossing the divide between what are now the states of Montana and Idaho, he continued to the Henry's Fork of the Snake River.

Ross returned to Flathead Fort in November of that same year and left in the spring of 1825. During the time he was trading out of Flathead Post he certainly would have acquired a great deal of knowledge about the region. On the following map note Ross showed Howse House to be almost due north of Flathead Fort near the shores of Flathead Lake.

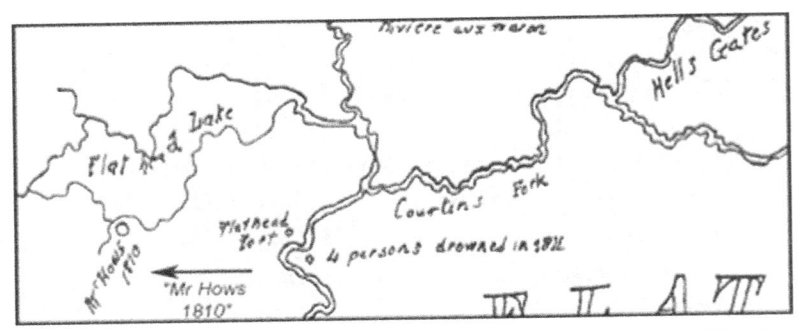

This portion of an 1849 map drawn by Alexander Ross shows the reported location of "Mr Hows 1810", an obvious reference to Howse House. North is to the left. The location of other important places and events also appear on the map.

James Bird noted in his journal entry on February 17, 1811, that:

> "Two Canadians Arrived at our Neighbors who brought us letters from Acton House. From these letters and from the information of our Neighbor, I am Acquainted that two of the NWt. Coy. Clerks Arrived at Acton House the 23rd Jany. From the place at which Mr. Howes is Wintering which they left on the 12th Decr., they have brought no letters fm. Mr. Howes: but we have the Satisfaction of Knowing that he reached the place of his Destination in safety and that they left him & Party in good health."

The reasoning used by B/W to place Howse House at Lake Pend Oreille was that many lakes in the region were referred to by more than one name. Thompson, however, referred to lakes in the region by more specific names. Those are the names that would have been used by McMillan and others familiar with the area. There is no question in my mind that when the reference was to Flathead Lake, there seems to be little reason to doubt it could only be Thompson's Saleesh Lake as shown on his maps.

In addition, early maps including those of Aaron Arrowsmith showed Howse House to be located in the Flathead Valley, not on the Clark's Fork (Saleesh) River.

Some posts operated only a short period of time for one or more reasons. The competition may have been too great, the location may have turned out to be inferior to that of the competition or they may have been re-located in order to gain a better proximity to major Indian roads or crossroads.

Years of experience in the fur trade industry taught both HBC and NWCo traders that when making a decision about where to build a post the most important criteria were location, location, location. It often made the difference between success and failure. Security also ranked high on the list of requirements. The location had to be defensible. In some cases that required a palisade surrounding the trading house.

A post had to be located where there was an adequate, dependable water supply and pasture for horses. Being close to a navigable river was very important since canoes were often the major means of getting trade items and supplies to western posts and furs and pemmican to company supply depots in Canada.

A Beaver Hunter

Ashley Creek Locale

One of the *least-likely* places selected by an experienced trader like Joseph Howse for building a trading post would have been on swampy ground. That factor would have been very important when it came to selecting the spot for Howse House.

The following General Land Office map (GLO) dated 1872, shows that during the period when the area was first mapped and opened to homesteading, extensive lands in the Flathead River floodplain were considered to be swamplands. As such they were not considered to be useful for much.

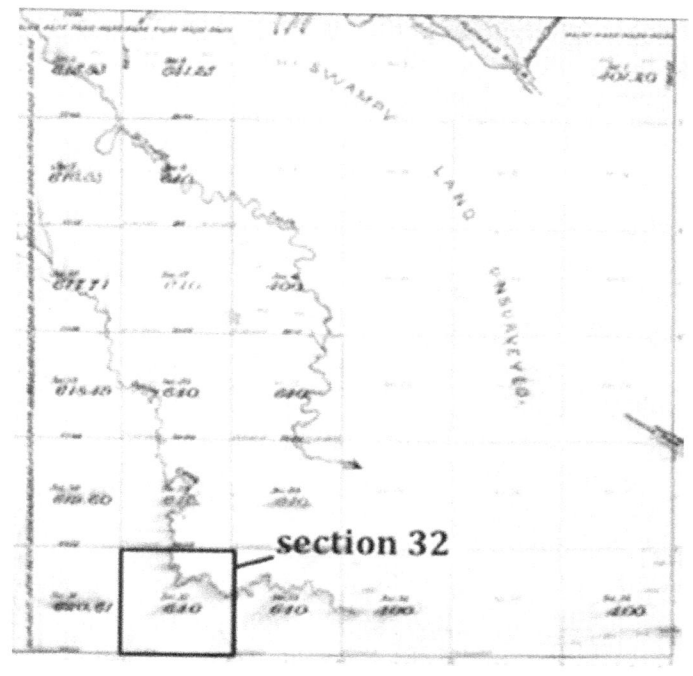

Note the "Swampy Land" notation on this 1872 GLO map. Such land would not be a desirable site for building a post. Also note the Ashley Creek channel ties in closely to the interior quarter section lines. Compare this to the 1892 GLO shown on page 32.

On this map, the Flathead River is found in the upper right and the Stillwater River is located between the Flathead River and Ashley Creek, which flows through Section 32.

The field surveyor classified the floodplain east of Section 32, the presumed location of Howes House, to be "swampy". It had been more than 60 years between the time that Howse House was constructed and this early map was drawn. I think it is safe to say that the *character* of the land had probably not changed much during that period of time. A later map shows the ground to be "agricultural".

It is interesting and important to note the mapped locations of the Ashley Creek channel through Section 32. Although twenty years older, the 1872 map was actually more accurate that the one drawn in 1892. The following comparison shows the actual channel location on satellite imagery. Compare the differences between that GLO map and the imagery of Section 32.

The swampy area probably froze when winter set in allowing Indian bands to camp there during the cold winter months. This was the time of the years when beaver pelts were in their prime. Beaver hunters from throughout the Flathead Valley and the surrounding mountains would arrive at the nearest trading post where the valuable pelts could be traded for blankets, cooking utensils, axes, guns, iron arrowheads, knives, beads and all kinds of other useful and enticing items.

When spring arrived, the unbridled, torrents of cold water from the deep, melting, mountain snow rushed into the valley spreading across the valley floor. Thawing of the frozen swamp combined with the flooding river quickly flooded the flats creating a swampy bog. By mid-summer much of the area would be dry again. Lush green grass to feed the horses covered the swamp and it was again possible to camp along the river where fish, whitetail deer, geese and other water birds were abundant.

Swamps and floodplains were not the place to build a trading post. Howse was an experienced HBC trader. It is highly likely that he selected a high and dry spot near one or more major Indian roads where grazing for horses was readily available; not in a spot where the post could be washed away in a spring flood or where knee-deep mud prevented hunters easy access to the post's trading room.

Mandatory criteria in locating a trading house were that it be on or near major Indian roads and near winter encampment sites. These criteria were clearly followed by Howse. The 1892 GLO map below shows the wagon road to the town of Demersville was located along an old Indian road. Note that it runs very close to the point where I believe Howse House was located. Another road was about half a mile west of the post site (see B/W page 115 for trail locations).

Using today's technology we know the Ashley Creek channel on the GLO was not correct. The GLO was sketched with a straight line from where it entered Section 32 to the point where it exited the section on the east line. The satellite photo shows the channel flowed through the northwest corner of the Eugene McCarthy homestead. A comparison of the two below illustrates how an error that escapes recognition can often make a great deal of difference when drawing conclusions. I apologize for the slight difference in scale.

This 1892 GLO interpretation of where the Ashley Creek channel enters and exits Section 32 are almost a straight-line connection between the established quarter section corners. Comparing this map with the 1872 map on page 29 indicates that the 1892 map was likely a re-survey of only the exterior section lines, did not include the interior lines, and thus resulted in an error as to the correct location of the Ashley Creek channel. Compare that to the same area from recent satellite imagery.

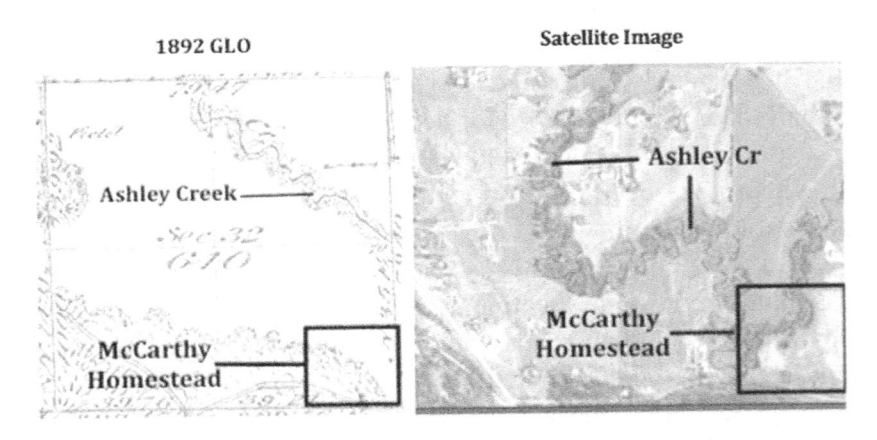

1892 GLO   **Satellite Image**

GLO maps were aimed at producing legal, government surveys. In simple terms, on the ground, surveyors set out from a monumented corner that was used as a control. From there they surveyed and chained the exterior lines to produce sections that were more or less one mile on each of the four sides and contained 640 acres. A "chain" is equal to 66 feet. The above maps show that the GLO's were not very accurate unless both exterior *and* interior lines were run and used to "connect the dots" where a feature such as the meandering Ashley Creek channel was involved.

For example, when a surveyor encountered a road or trail along a line being surveyed he would note its distance from the corner to the crossing and often draw a line indicating where the road crossed. Interior detail in a section was mostly based on best guesses and "connecting the dots" so to speak. Ashley Creek is a great example. On the map above, it is shown flowing across the NE quarter of Section 32, when in fact it continued flowing in a southerly direction to the foothills then doubled back to the northeast. Since those two points were where the surveyors crossed the creek they simply connected the dots

.The following Google contour map shows the location of the McCarthy homesteads referenced by Emma Ingalls. I am convinced that somewhere within the circle is where Howse House was located.

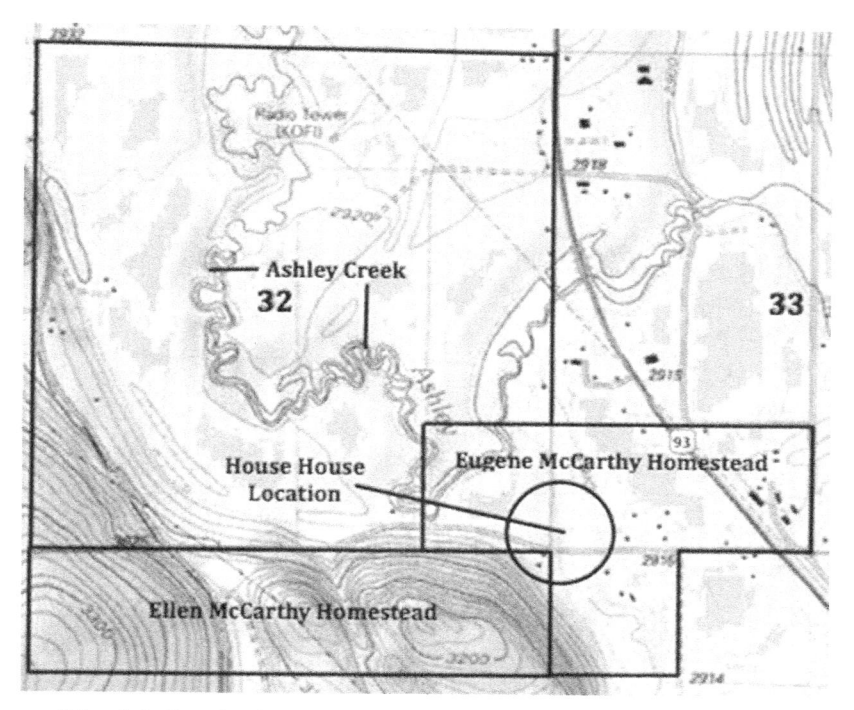

The McCarthy homesteads are shown here. Ashley Creek was named for Joe Ashley who first settled here in 1857, on what later became the Eugene McCarthy place.

One of three things likely occurred at the post following its abandonment by Howse. All are typical of what might be the fate of abandoned structures at the time. The first is that perhaps an independent beaver hunter/trader—a "freeman"—moved in and continued trading in the area. The second is that an abandoned log structure was often torn apart and used for firewood by parties camped in the area. The third is that someone moved into it instead of building a new cabin elsewhere.

MUZZLE LOADING PISTOL BARREL
FROM WESTSIDE OF ASHLEY CREEK

(Sketch is not to scale)

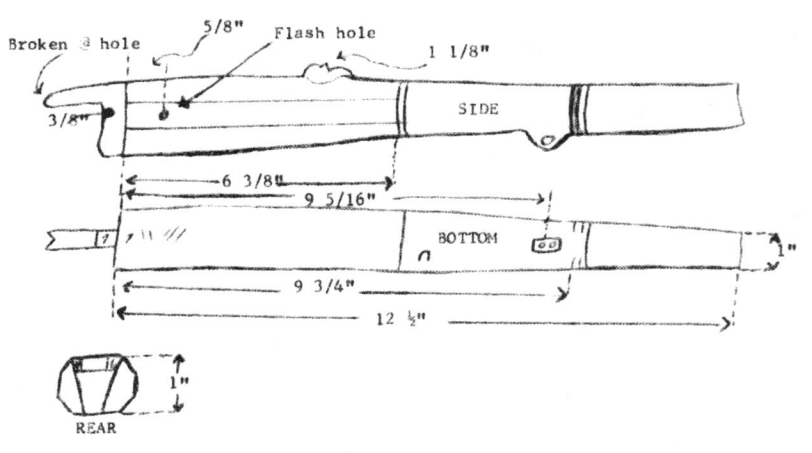

DESCRIPTION

Weight: about 1 3/4 lbs.
Caliber: 0.562" or larger
A #4 is located on the left side of the barrel 1 3/8" toward the muzzle on the 10 o'clock octagon. Also there is a letter C on the bottom of the round portion of the barrel 6 7/8" from the rear and a pair of matched 7s on the plug and on the barrel.

Peninsula Locale

Some proponents of the peninsula location of Howse House base their conclusion on the idea that since canoes were a principal means of transportation during the fur trade period that a riverside trading post made good sense. And, it would have had the floodplain not been prone to flooding in the years prior to the construction of Hungry Horse Dam on the south fork of the Flathead River. An experienced trader would not have built a log building on a spot that was under water or swampy a large part of the year. Joseph Howse would have known better.

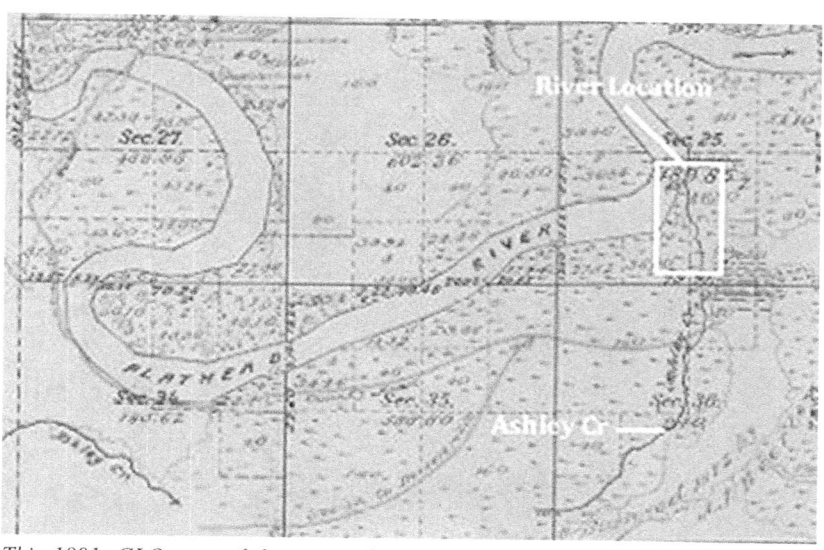

This 1891, GLO map of the peninsula at the confluence of Ashley Creek and the Flathead River in the NESW quarter of Section 25, T28N R21W shows the approximate location where some think the trading post was likely located.

An old pistol barrel, supposedly of French origin, was found on the peninsula between Ashley Creek and the Flathead River. B/W drawing of the artifact is shown on previous page. Some have viewed the barrel as evidence that Howse House was located near where barrel was found. During the past 200 years a significant number of guns were undoubtedly lost west of the Rocky Mountains. An item such as this pistol cannot be directly associated with Howse House unless it were to appear on a cargo manifest, its loss clearly mentioned in a journal entry, or some other mention that would tie it to Howse and that precise location. In my opinion, a conclusion that this pistol was actually associated with Howse House can only be reached by means of speculation.

This satellite image shows the peninsula where Ashley Creek empties into the Flathead River. The location of the "present log house" and the spot where the old pistol barrel was found are based on commentary and maps from the B/W paper found in Appendix A.

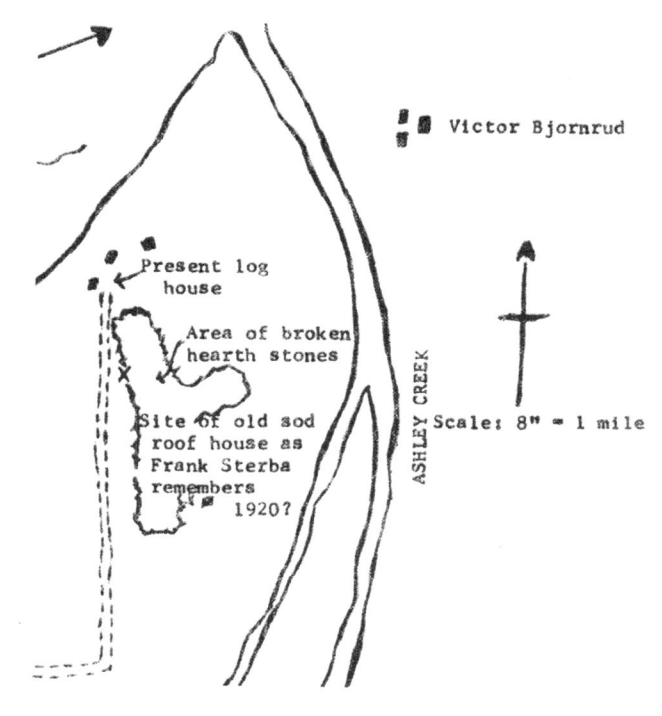

This drawing of the peninsula from the B/W report shows the location and condition of several old buildings at the time of their visit to the site. An experienced company man, like Howse, would not have built a post on a site that prone to flooding almost every year. Wood used in such building that was in contact with the ground or floodwaters would have quickly rotted away.

Braunberger and White also concluded that:

"Despite the fact that the Arrowsmiths placed Howse House on their map at the head of present Flathead Lake, the evidence weighs more heavily in favor of Howse House having been established along the present Clark Fork River (above present

> Lake Pend Oreille)... The report of McMillan and Montour would point to a Clark Fork River location, as well as Thompson's reference to the H.B. 'in the lake'."

I disagree. Here is why. The B/W manuscript includes the following:

> "There remains but scant evidence from which to speculate upon the actual journey taken by Joseph Howse beyond the southernmost point of latitude on the Kootenai River. The maps Howse made in 1812 and 1815 and consigned to the mapmaker, Aaron Arrowsmith, were never returned to the Hudson's Bay Co. All of the Arrowsmith maps were sold at public auction on 28 July 1874 following the death of John Arrowsmith in May, 1873."

The Arrowsmith's were considered to be among the most revered mapmakers of the period. If Arrowsmith had the Howse maps as researched and reported by B/W, then it would seem logical that Arrowsmith's map—and thus the Howse House location on it—would have been relatively accurate.

Both of Thompson's posts on the Saleesh River, Kullyspel House and Saleesh House, were actively engaged in trading during the time when Howse built the HBC post. He was aware that Howse had been sent to find out what the NWCo was up to and would have been intimately aware of where the HBC post was, or had been, located. That would have been especially true had it, indeed, been in the Clark's Fork Valley. That being the case, there would undoubtedly have been reported contact between the two companies.

Neither Thompson, in his often-detailed journal (and later in his narrative), nor his men, ever mentioned direct contact. The only exception even close would be, as the B/W manuscript points out:

> *"There remains the report of McMillan and Montour from the Alexander Henry journals upon their return to Rocky Mountain House wherein they state that: 'the H. B. Co is also settled on Flat Head Lake.'"*

James McMillan, accompanied by Nicholas Montour, was dispatched to find out where Howse House was located and how the Howse brigade was doing. In fact, as discussed earlier, Howse did quite well during his first season. According to the HBC archives, Montour was stationed at Fort Vermilion on the Saskatchewan River at the time. In 1811, he was listed as being on the "Columbia" and in 1812–1813, at Thompson's Spokan House operation near Spokane, WA.

McMillan undoubtedly reported to Thompson what he and Montour had observed of the HBC operation. Had it been Kullyspel (today's Pend Oreille) Lake, McMillan would have known exactly where the Howse post was located since both Kullyspel House and Saleesh House were actively engaged in trading at the time. In the summer of 1811, Thompson ordered the closing of Kullyspel House following his delayed return from Canada due to the Piegan blockade on the Howse Pass trail.

1859

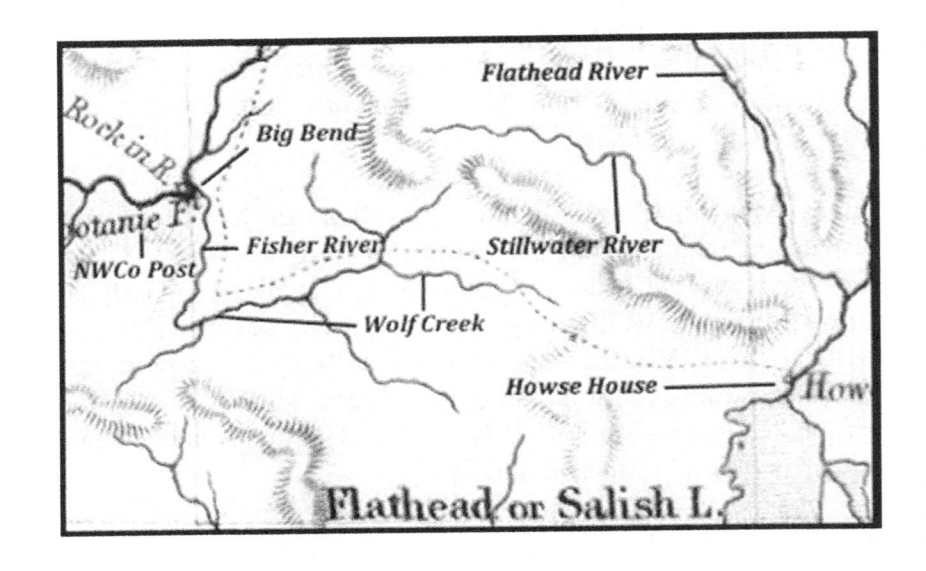

The Route

The Indian road that crossed from McGillivray's River to the Saleesh River did follow the Rock (Fisher) River to the Thompson Lakes, down Thompson River to the Little Thompson where it swung to the southeast then even more southerly down the Buffalo Bill Creek drainage to the Saleesh River at Weeksville.

The Dunn Creek road led to the Flathead Valley. A fork of the Rock River road led to the Saleesh River, another to the Flathead Valley. Since we do not have much to go on, the fact that the post appeared on several early maps lends credence to the Flathead Lake location. Howse' reference to the "southern most bight" of the river and of "Poile de Castor" support the idea that he was either misguided or missed the Indian road that would have led him to the Saleesh River.

An old road up the Dunn Creek drainage as well as the one along Wolf Creek used by Thompson during his travel from the Saleesh House site to the Big Bend in 1809, both led into the Flathead Valley.

Mark and I were in agreement, that Howse was likely misled by his Indian guide. The trail Howse should have taken led south up Fisher (Rock) River and then down Thompson River to Weeksville on the Clark's Fork (Saleesh) River where Thompson's posts were located.

Instead Howse probably followed the Dunn Creek trail in the area of the Poil de Castor area, crossed over to Wolf Creek, which he followed to the Flathead Valley located to the east.

The 1859 map on page 40 shows the trail from Howse House to the Big Bend on the Kootanai (McGillivray's) River following the Wolf Creek route. It is highly likely the 1859 trail was the same Indian road that existed in 1810.

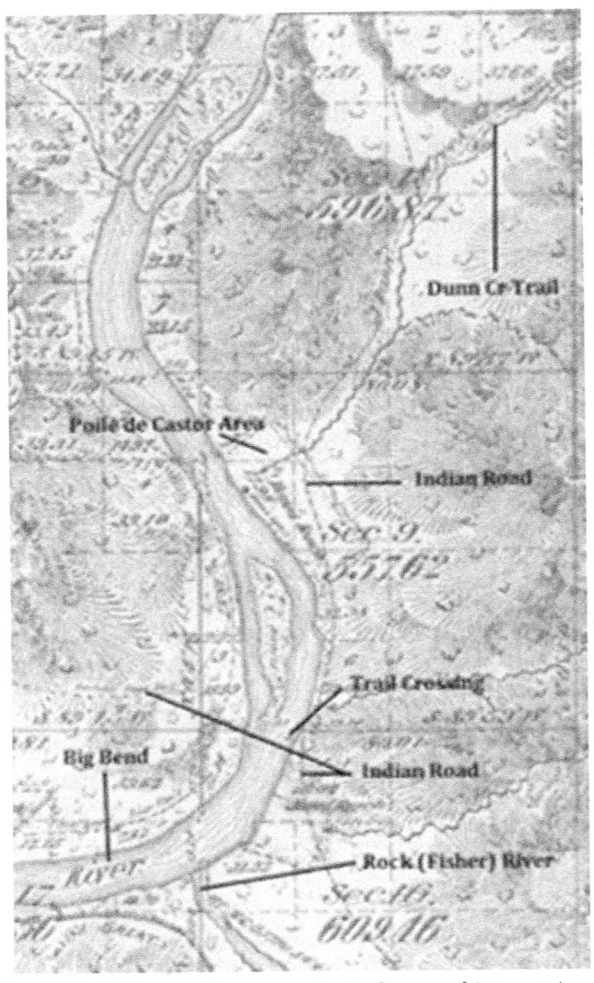

This 1898 GLO map shows a major Indian road intersection at the Big Bend of McGillivray's (Kootenai) River. A road to the west crossed the river at the "trail crossing" then led to the Lake Indian Road near Bonners Ferry, ID, which Thompson and his brigade had followed to get from the Kootenai River to the Saleesh River at Lake Pend Oreille.

One Final Mystery

There is one other little-known, unanswered mystery about Howse exploration and mapping during 1810–1811 we should at least mention. In the Columbia Gorge Interpretive Center, located near Stevenson, WA, there is a carved rock allegedly—and likely—connected to the HBC. Following is information about the rock was taken from the Museum's website. The rock is pictured under the heading *"No Longer a Mystery. The initials speak for themselves"*:

> *"In 1923 a road crew pulled the Skamania County "mystery rock" from the Government Slide near the Cascades. For decades the stone stood near the entrance to Stevenson High School. Its inscription -- "HBCo[y] 1811" -- remained unexplained. Retired attorney David Swart, a man deeply interested in the early fur trade, believes the stone marks the expedition of Joseph Howse. In 1810–11 Howse led a party of seventeen men across the Rockies and descended the Columbia River to at least this point more than six months before the well-known explorer David Thompson. Unfortunately neither Howse's journal nor letters written by him were preserved; Thompson's survived."*

In my view, it takes a rather large leap to associate the rock with Joseph Howse. Here is why. First, as far as I know, there is no historical evidence of any kind supporting—or even suggesting—that Howse made a trip *down* the lower Columbia. In order to get to the Flathead valley he had to travel *up the river* to its headwaters in order to cross the divided between the Columbia River and today's Kootenai River. There is little doubt about that part of his journey.

Howse did not arrive in the Flathead Valley until late summer or early fall of 1810. Once there, his attention had to be focused on finding a good location for his post, building it and contacting Indian bands in the area about his desire to trade furs for goods. During that winter he made a trip to the *"head branches of ye Missouri"* returning to his post on February 9, 1811, then back to the Upper Saskatchewan in the summer.

Could he have made a trip down the lower Columbia during that time? The answer is yes—probably. Did he? Due to a lack of historical evidence, my guess is he did not. HBC records indicate Howse drafted a map of the Columbia River in 1815. But the question remains—was that map only of the upper reaches or did it include portions of the lower Columbia?

HBC carved in rock. Photo by Lyn Topinka, 2011

Conclusion

I believe the best historic evidence available, points to Howse House having most likely been located in the SESE Section 32, T28N, R21W, in the SWSW section 33 or the NWNW Section 4 T27N, R21W along today's Rocky Cliff Drive as shown on the map on page 33.

Based on interpretation of available evidence, I feel confident that if we have not identified the true location of Howse House, we are very, very close.

My conclusions are based on the following facts:

– Although they have evidently disappeared, Joseph Howse sketched some maps when he was there. He did not have surveying equipment with him in which case they would have shown only a relative position to his surroundings and landmarks.

– The maps were given to mapmaker Aaron Arrowsmith by the HBC but were never returned. The Arrowsmith collection of maps was sold in 1873. As far as we know, they have never resurfaced.

– Since Arrowsmith had the maps and sketches drawn by Howse, it seems logical to assume that his maps would have shown the relative location of Howse House to Saleesh (Flathead) Lake. I believe they did.

– Later maps, such as the one Alexander Ross drew in 1849, continued to show Howse House located near the north end of Saleesh *Lake*; not on the Saleesh *River*.

– Joseph Howse provided key information about his route to the place where he built his trading post: *"Descended the most Southly Bight of it* [the Big Bend of the Kootenai River] *crossed portage poil de Custer to Flathead River* [above the Lake; see Arrowsmith] *where we built."* Based on what we know about the location of Poile de Castor, there are at least two routes to the Flathead Valley he could have followed.

– Emma Ingalls written comments strongly suggest the location as being on the old McCarthy homestead and that *"where this historic building once stood Mr. Griffith has a grain field."* I have not been able to locate Griffith's grain field, but based on Ingall's comment, there is little doubt the field was on the McCarthy homestead.

– The location was near the confluence of several major Indian roads. Such roads in the area generally followed the "foothills" rather than crossing the extensive swampy lands on the floodplain.

– The post would not have been built on "swampy" land due to arguments discussed earlier.

– Neither James McMillan nor David Thompson ever noted it being on the Saleesh (Clark's Fork) River. The NWCo's trading posts, (Kullyspel House and Saleesh House), were both located at strategic points on the river. Both men were very familiar with Lake Pend Oreille from the mouth of the Saleesh River to its outlet on the west side. Had Howse House been located in their trading area on the Saleesh and Flathead Rivers, I believe Thompson would have included, in his journal or narrative, at least a short reference to information of this importance.

– It is possible that I, like others before me, have missed something significant in our research somewhere along the way. If there are clear errors in this manuscript, I hope those more knowledgeable than I will come forward with historical evidence that will shine additional light on this subject that will be useful in resolving this 100 year-old controversy.

A Hudson's Bay Trading House by C. W. Jefferys

References Cited

Coues, Elliot – *The Manuscript Journals of Alexander Henry and of David Thompson, 1799-1814*, Volumes I and II, Ross & Haines, Inc., Minneapolis, Minnesota, 1897, 1,500 copies reprinted in 1965.

Johns, Sam E., 1943 – *Early Settlers of the Upper Flathead Valley*

White, Catherine, 1950 – David Thompson, *Journals Relating to Montana and Adjacent Regions, 1808-1812.*

White, Mark - *1810-1811 Howse House on the Flathead River North of Flathead Lake*

White, Thain and Braunberger, A. B., 1962 – *Howse House, An Examination of the Historical and Archeological Study,* 90 p. illus., maps. typescript (photocopy)

Wolfart, H. Christoph – *Dictionary of Canadian Biography Online*, University of Toronto/Université Laval

APPENDIX

HOWSE HOUSE

AN HISTORIC AND ARCHAELOGICAL STUDY

BARRY BRAUNBERGER & THAIN WHITE
1962

PLEASE NOTE: *Since we did not have access to the original photographs in this paper, we chose not to include them due to quality issues associated with using a copy from the typewritten original. We included the those illustrations and maps that could be reproduced. The text is, in most cases, self-explanatory. We have also chosen to italicize quotes.*

Carl W. Haywood
Author

CONTENTS

PART I

Joseph Howse's journeys west of the Rocky Mountains; Hudson's Bay Posts along the Saskatchewan; James Bird's correspondence and arrangements for Howse's expedition; Alexander Henry, the Younger's description of Hudson's Bay Company departure; Hudson's Bay Co. account book listing of men under contract; activities of Jaco Finlay and James McMillan; Bird's correspondence from Edmonton House Journal; J. B. Tyrrell's observations; extracts from Alexander Ross's narratives; extracts of letters from Howse and Cook to (Joseph) Colen; Colen to Banks; Semple to Douglas, and Howse to Simpson.

PART II

Discussion of the name Flathead, in relation to rivers and streams in Montana; Flathead Indian travel routes from the Bitterroot valley; Howse's probable route southward from the Kootenai River; Lake Pend d'Oreille versus Flathead Lake as the site for Howse House, and conclusion.

PART III

A discussion with illustrations of a total of ten Archaeological sites personally visited by the authors within the Flathead valley; listing artifacts obtained, concentration points and general information.

APPENDICES

A. Derivation of the name, Kutenai 114
B. Indian Roads in the Early Days 114
C. Lake & River transportation 121
D. Data on Scarred Trees, Ashley Creek area 124
E. Kutenai Informant sites 125
F. Discussion of booklet by Grace Flandrau 129
G. Buildings and methods of construction 131
H. Information from local residents 132

ILLUSTRATIONS

Muzzle loading pistol barrel 96
Net sinker 100
Projectile points 102
Hatchet tracing 102
Cross-section of peeled tree 123

MAPS

Tracing from World Aeronautical Chart showing the Flathead, Clark Fork and Pond d' Oreille Rivers .... Not Shown
Archaeological sites- Flathead Lake Region, up to 1953 ..92
Field surveys by authors in Flathead valley 93
West Side Ashley Creek site 94
H. D. Lockhart site 103
East Side Ashley Creek site 104
Pete Brosten site 105
Brosten & Therriault Ferry sites 108
T. W. Wagner site 110

MAPS CONTINUED

Hetland Springs site 113
Indian Trails- Flathead Lake Region 115
Indian Trails & winter camping areas 120
Great Northern RR to Somers 130
Enlargement of Flathead Lake portion of Arrowsmith map (1795- 1819) page 134

PREFACE

The material compiled and presented within this paper was the result of four years of research, study and investigation concerning the life and travels of one Joseph Howse, in his service with the Hudson's Bay Company.

The primary intent of this study was to locate, or at best approximate, the site of the winter camp established by Howse during the winter of 1810-11 and referred to as "Hose House" in the company records.

Howse was by no means one of the most outstanding men of his era, but he undoubtedly accomplished an outstanding and singular achievement in the performance of his duties. He had the enviable distinction of living in a fast moving and colorful generation. A generation shadowed by constant insecurity, but one which allowed unrestrained freedom to those accepting the challenge of the wilderness. All of this tempting and abundant opportunity vanished in the early 1840's when the market for beaver dwindled and vanished.

The fur trader was a highly individualistic person who led a lonely but most self-sufficient life. He was constantly subjected to, and readily accepted, many and unique experiences in the remoteness of his hunting grounds; he was more than willing to accept great land distances to achieve his prime purpose, but he was definitely not a pioneer consciously clearing the way for others to follow—he was only hunting beaver and unknowingly exploiting the remote wilderness for which he gave little in return.

As was most often the case, the fur trader was not an articulate individual, and cared even less to burden himself with a notebook and pen with which to record his travels for the sake of posterity. It is this lack of record, the void, and the uncertainty of purpose, that provides the incentive and the necessary stimulus to the historian- and for this manuscript.

Joseph Howse, is properly given credit, the singular honor, of having established the first Hudson's Bay Post west of the Rocky Mountains. The biography of Joseph Howse as contained in the Colin Robertson's Correspondence Book reveals that Howse lived to be 78 years of age, was a native of Cirencester, Gloucestershire, England. Here he was born, raised, and lived most of his life with the exception of a rigorous twenty years in the Canadian wilderness in the service of the Hudson's Bay Co.

Howse began his service with the Hudson's Bay Co. as a writer at the age of 21 in the year of 1795, and his advancements during the following eight years proved him to be an alert and comprehending individual, and a literate one as well. There would be small reason to doubt that he possessed these qualities; not only in regard to his record in Canada, but because of his painstaking endeavor to preserve the grammar of the Cree language by a publication in 1844 (this publication was accompanied by an analysis of the Chipewyan dialect).

Although Howse left active service of the Hudson's Bay Co. in the fall of 1815 to return to England, it is not clear from his biography whether he left the Hudson's Bay Co. of his own volition or was discharged by his superiors. The conditions concerning his departure for England are open to casual speculation; as the record shows Howse was, to some extent, discredited in the correspondence of his superiors.

H. B. Auld, one of Howse's superiors at York Factory, accused Howse of aiding and intensifying the Flathead-Piegan hostilities during the summer of 1810 by giving the Flathead Indians firearms. This accusation, directed at Howse, would seem to be premature and quite unjustified, as Howse did not depart on his expedition to the Columbia until June of 1810, and if he did possess firearms for trade with the Indians, it is unlikely this equipment reached the hands of the Flatheads in sufficient time to assist this tribe in their encounter with the Piegans.

Defense of Howse against this unfair accusation, is contained in the Thompson Narratives where David Thompson in his discussion of the Saleesh-Piegan battle of July 1810 states:

> 'The Saleesh Indians during the winter [1809–10] had traded upwards of twenty guns from me, with several hundreds of iron arrow heads, with which they thought themselves a fair match for the Piegan Indians in the battle on the Plains."

In *A History of Montana*, 1957, Chapter on the Fur Trade, Paul C. Phillips states:

> "McDonald equipped the Indians with firearms and made plans to accompany them across the mountains to the buffalo country. In July 1810, this expedition started out. It passed by Flathead Lake [Salish] and then by a 'wide defile of easy passage' probably the Marias Pass, and McDonald and his companios were doubtless the first white men to pass through it."

The *Daily Inter Lake* of Kalispell, Mont. in its progress edition of Sunday April 24, 1960, carried an article on page D10, *Marias Pass Played Important Part in Early Park Exploring*. In part the article states essentially the same information as above, but goes on to add: *"At a spot believed to be just below the old railroad siding of Skyland on Bear Creek, the party was attacked by 170 Piegans and a furious battle followed."*

Despite the above information the authors in compiling information for Appendix B (*Indian Roads of the Flathead Lake Region*) of this paper would contend that the two main Indian east-west travel routes in this area were Aeneas Pass and present day Logan Pass. Aeneas Pass being the closer to Flathead (Salish) Lake was probably the "wide defile of easy passage."

Another effort to discredit Howse appears in a portion of a letter from (Colin) Robertson directed to (Mr.) Irving about the end of December of 1817. Robertson states:

> *"I only know him from a futile attempt to establish a post at Ile-a-la-Crosse, where a brave young man lost his life and his shameful retreat from that place in the spring; his name I believe is Howse."*

The material presented in this paper is garnered reluctantly from both second and third-hand information, as there are no specific journals or records available for reference.

Much of the Indian information contained in this paper has never before appeared in writing and we feel it is a reasonable premise to assume the information is authentic as it was obtained directly from our Indian informants.

Included to supplement the basic presentation are eight appendices. Herein is entered a discussion of the Indian Roads in the Flathead Lake area, archaeological sites, information from local residents, from Indian informants, data on scarred trees, etc.

A substantial portion of the presentation comes under the heading of, Archaeological Sites. This material is presented to supplement the surveyed and numbered archaeological sites as shown on the map of archaeological sites in the Flathead Lake region by listing and describing the sites we have personally visited and have made appropriate notations as to our observations.

In all, a total of eleven sites are listed under the archaeological section, and these areas of investigation are presented because the topography, the convergence of Indian Roads, the early settlements, reports of artifacts, and information from informants made these sites likely prospects for investigation.

ACKNOWLEDGEMENTS

We are deeply indebted to many people in the preparation of this paper; in particular we would like to give deserved recognition to Dr. Merrill Burlingame of Montana State College for his many and valuable suggestions; to Mr. Hugh A. Deapsey, Archivist, of the Glenbow Foundation, for his most helpful suggestion in directing us to publications, both new and old, which helped immeasurable in our search for information. We are under obligation to Mrs. Virginia A. Deeter and Mr. Bruce Schuyler of the Bureau of Land Management in Billings for their kindness and attention to our many and detailed requests for information from their office. We are indebted also to Mrs. Emmett Avery, Archivist, of Washington State College for preparing numerous photo static copies of documents; to Mr. Pierre Nove and Mrs. Leslie Sterling (of Bigfork and Kalispell) for their valuable assistance in the translation of French phrases and idioms to the English language. Our further obligations are extended to Mrs. Florence I. Vinal, Circulation Librarian, of the Montana State Library Extension Commission in Missoula; to Mrs. Jeane Sturidivant, Librarian, of the Carnegle Public Library in Kalispell; to Mr. T. R. McCloy, Librarian, of the Glenbow Foundation in Calgary; to Mr. Vernon D. Hawley, Fur Resources Biologist, State of Montana Fish & Game Department; to Mr. Ed Sager, Resident Gene Warden, Kalispell.

We are most grateful to the services and information provided by Mr. John Hove, of Francis Edwards Ltd. of London, England; to Dr. W. Kaye Lamb, Dominion Archivist, of the Public Archives of Canada for information and for preparing photo static copies of the Arrowsmith maps.

Our acknowledgements extend to both Frank and Ed Trippet of Trippet's Printing Co. who were most helpful and considerate in providing back copies of the Kalispell Times for our use.

We are deeply indebted to Miss Alice M. Johnson, Archivist, of the Hudson's Bay Co. in London, England. Miss Johnson not only directed the search for the information we most desired, but was most gracious in acknowledging our correspondence in request for information from the Hudson's Bay Archives.

In addition, we are under obligation to many people in the preparation of this paper; to land owners, the Indians who assisted us, and too many others who have taken a genuine interest in our work. Unfortunately, it is impossible to give individual acknowledgement to everyone, but we are most sensible to everyone who gave of their time and knowledge.

    Kalispell, Montana A. Barry Braunberge
    Lakeside, Montana Thain White

JOSEPH HOWSE'S JOURNEYS
PART I

In the years immediately preceding Joseph Howse's crossing of the Rocky Mountains, the Hudson's Bay Company's posts along the Saskatchewan River, in what was called the Saskatchewan district, and were in charge of James Bird. His headquarters were at Edmonton House on the North Saskatchewan River, very near to Fort Augustus of the North West Company. At that time the rival posts were on the site of the present city of Edmonton, Alberta.

At the end of trading season 1807–08 James Bird, accompanied by Peter Fidler and Joseph Howse who had spent the season at subordinate posts in the district, took their fur returns to York Factory, the Company's depot on Hudson Bay. They arrived at York on 29 June 1808 and left on their return to the Saskatchewan district with the trading goods for outfit 1808–09 on 11 July following.

Alexander Henry, the Younger, in his journals first mentions Joseph Howse under the journal entry of Sunday, 28 August 1808. In August of 1808 Henry was on the lower Saskatchewan in the vicinity of Birch Island preceding westward up the river, and Howse was on his return journey en route from York Factory to Carlton House. Henry's 28 August journal entry states:

> "...We passed the remains of an old establishment, abandoned many years ago. At dark we overtook Mr. Howse of the H. B. Co. from York Factory, bound to the South Branch of the Saskatchewan, with a large boat containing about 20 pieces, worked by eight Orkney men."

Henry continues the following day (29 August):

> *"Before day we were on the water, leaving our H. B. Co. gentlemen still sound asleep."*

While he was at York Factory Bird evidently had some discussions with his superior officer, John McNab, about the possibility of sending an expedition to the west of the Rocky Mountains, for in a letter written to McNab from Oxford House on 30 July 1808 he remarked: (H.B.C. Arch. B. 239/2/114 - Elliot Coues, editor, *New light on the Early History of the Greater Northwest,* New York, 1897:

> *"I arrived here with Messrs. Howse & Fidler in three Canoes the 25th ins...*

The expedition across the Rocky Mountain does not seem to be quite relinguished by the N.W.Co. though their success last year appears not to have been encouraging.

A Canoe manned by five Iroquois & carrying sixteen pieces of goods, they have again sent there, principally as it appears with a new of drawing Beaver they have in vain hoped to receive from the Natives.

I shall expect to hear by the fall Canoe your wishes with regard to the west side of the Mountain in every point of view; it must always however be understood, that a few extra men will be required for such an undertaking should it appear eligible..."

To this McNab replied from York Factory on 30 August 1808:

> *"I received your letter of the 30th July...*
> *You know my sentiments respecting the Rocky Mountain accord with yours as our wishes depend on the aid of extra men they must remain unfulfilled..."*

According to H. B. records, Howse wintered at Carlton House on the South Branch of the Saskatchewan River during 1808-09 and in May 1809 he was en route to Edmonton House where he was to take charge while James Bird paid his annual visit to York Factory. Bird wrote from Oxford House to Howse on 25 July:

> "You will herewith receive an adequate supply of trading goods & a sufficient number of men for every requisite undertaking..."

And added:

> "As you are acquainted with Nature & Extent of the honorable. Company's concerns at Edmonton, & as I have already given you every information that I could conceive to be requisite, I shall add nothing further on the subject, than to say that I have the utmost confidence in your Zeal and Ability, and to wish that your exertions may meet with the success I know they will meet..."

James Bird's remarks suggest that Joseph Howse was to do something more than the usual routine duty of looking after the business of Edmonton House during the summer of 1809, but of this we have no confirmation because the most likely sources of information for trading season 1809–10 are missing from the H.B.Company's archives and no mention of Howse's movements can be found in the Journals of Alexander Henry the Younger, and David Thompson of the Northwest Company.

According to both Elliott Coues and J.B.Tyrrell, David Thompson met Joseph Howse in the summer of 1809 returning from an exploring journey into or towards the Rocky Mountains. In the York Factory Account Books for outfit 1809–10 Joseph Howse was listed as being stationed at Edmonton House during that trading season, and his hope that his salary would be raised from the amount of 65 to 80 per annum was recorded. The recommendation reading:

> "the readiness with which this Gentn. undertook the expedition across the Rocky Mountain merits some attention."

must have been added by James Bird, and this recommendation, in turn confirms David Thompsons statement that Howse had made a journey into the Rocky Mountains in the summer of 1809.

According to Elliot Coues:

> "A Mr. Howse is noted by Thompson at Fort Vermillion July 18, 1809, and Thompson met him [Howse] in the Rocky Mts. Near the head of the N. Saskatchewan Aug 8th of that year, with one man [White] and an Indian Mr. House left Fort Vermillion for Fort Augustus 23 September 1809."

This statement is corroborated by an entry in the Henry journals of the same date (23 September) stating:

> "Mr. House set off for Fort Augustus on horseback."

The occasion for Henry to further note the activities of Joseph Howse is on 18 May 1810 while Henry is still in residence at Fort Vermillion. on this date he mentions:

> "Mr. House arrived from Terre Blanche in the boat Mr. Hallett went up in."

Again, because of the most likely sources of information are missing from the H.B. Company's archives, reference is made to Alexander Henrys journal for the exact date in 1810 on which Howse began the journey that was to take him into present-day Montana. According to Henry, the Hudson's Bay Company men began their journey on 19 June in two canoes

> "*for the Columbia, with nine men including the two Pacquins.*"

Henry continued:

> "*They embarked four rolls of tobacco, two kegs of high wine, powder, several bags of balls, a bag of shot, pemmican etc... 20th* [June] *Mr. House, Mr. John Parks, Willock with four Cree guides and hunters, the youngest of the Pacquins, and a number of houses, off by land; the whole H.B.Co. Columbia expedition consists of 17 persons, including the four Indians...*"

The party of 17 was thus made up of ten company employees, four Cree Indians and three members of the Pacquin family. The youngest refers to the last or youngest member of the family. Coues refers to David Thompson's journals of 23 June 1810 when Thompson was coming down the Saskatchewan River from the Columbia, and had preceded below old Fort Augustus.

> "*... we passed two H.B.Canoes, well arranged for the Columbia*".

The only direct record in the H.B. Archives of Howse's venture during 1810–11 is an account book headed:

"S[askatchewan] *F*[actory]
Columbia or Flathead River Accounts
1810–11"

and this makes no reference to the Pacquin family or the Cree Indians. The men under contract to the company who accompanied Joseph Howse were, according to this account book,

> John Ashburn
> William Bruse
> William Harper
> John Howrie
> Joseph Lewes
> John Morwick
> John Park
> William Taylor
> James Whitquay

Confirmation that these were the only men under contract to the company and in receipt of annual wages who were in "the Columbia" during 1810–11 can be found in the Edmonton House account book referred to in note 12 on page 10.

Other writers have speculated and even stated categorically that Jaco Finlay accompanied the Howse expedition to the Columbia as a freeman. The evidence does not support this contention. Rather, it would appear that Jaco Finlay was engaged by David Thompson and was assisting Finan McDonald with the building of Spokane House during the winter of 1810–11.

Alexander Henry at Terre Blanche House [Fort Augustus, no. 3] on the North Saskatchewan River learned on 22 October 1810 from the returning Cree, Mawtoose, who had started off with Howse, but was returning to look for his family, that the Hudson's Bay people had gone "to the old kootenay house" on the Columbia River.

While Howse made final preparations and ultimately got his Columbia expedition under way, James McMillan, of the N.W.Co whose travels in the succeeding months were to be intimately associated with those of Howse, was returning with David Thompson up the Kootenay River from Kullyspell House. On 14 June 1810 during the return trip McMillan was delegated by Thompson to attend to the packs together with Messrs. Method and Vandette at the kicking horse rapids on the Columbia River while Thompson, in obvious haste, crossed the mountains to the eastward.

The arrival of James McMillan from the Columbia was noted by Henry at Terre Blanche on 5 July 1810 and after four days of rest he was instructed to take off again and to watch the motions of the H. B ... when he began his journey on 9 July, McMillan was nineteen days behind the Howse party. It is apparent that McMillan caught up with the Howse expedition at the old Kootanay House and according to the returning Cree Mawkoose, the rival H.B. and N.W. parties had, at the time he left, been... "prevented from descending further by the Piegans and Fall Indians" who were watching the Northwest People on McGillivray's [Kootenay] River.

Meanwhile, James Bird had arrived at Carlton House en route to Edmonton House on 18 October 1810 and there entered in his journal:

> "Received also Letters from Mr. Howse and party who are well; and had Embarked safe on the Cootana Columbia River on their way to the Columbia..."

Bird reached Edmonton House on 31 October following and recorded in his journal on that day:

"Received Letters from Mr. Howse, dated Cootana River 20th Augt. 1810, in which he says that, having been informed by some Cootanahas that a Battle had been fought between a party of Flatt Head Indians, with whom a Mr. McDonald clerk to the N. Wt. Coy was in company and a party of Muddy river [Missouri] *Indians, in which the latter were defeated with the loss of 14 men killed* [according to Thompson's Narratives]; *and that the Muddy River Indians in consequence were laying in ambush to intercept him or any white man who might attempt to convey goods to the Flat Heads, he had determined on remaining some time at the place where this letter was dated to gain further intelligence. Mr. Thompson it appears after a settlement had been formed & goods left for the Muddy River Indians at Acton house set off privately with twenty men for a year's abode near the Columbia..."*

The following extracts from Bird's Edmonton House Journal also refer to David Thompson and Joseph Howse during the winter of 1810–1811:

Jan 14 *"it appears that Mr. Thomson is passing the winter close to the rocky mountain in one of the branches of the Athabaskow River, from where he expects to find a passage to some branch of the Columbia next spring; our men from Acton House met seven of his men on their provisions..."*

Feb. 17 *"Two Canadians arrived at our neighbors who brought us letters from Acton house. From these letters and from the information of our neighbor. I am acquainted that* two of the NWt. Coy. Clerks arrived at Acton house the 23rd Jany. From the place at which Mr. Howse is Wintering which they left on the 12th Dec. they have brought no letters from Mr. Howse; but we have the satisfaction of knowing that he reached the place of his Destination in safety, and that they left him & party in good health."

Mar. 25   "sent off two men to go to Acton House, and from these with horses & Pemican to meet Mr. Howse at the Cootanha River, on the west side of the Rocky Mountain..."

May 10   Extract from a letter to J.P. Pruden at Carlton House: "as it had become necessary to occupy Acton House during the summer and as Mr. Hallett Will in a short time leave the service, your presence is necessary here, for the managing the affairs of this place till Mr. House arrives..."

May 13   Men arrive from Acton House. Wm. Flett and four men remain there to pass the summer. "... A desire to conciliate the minds of the Indians, and to dispose them as much as in our power to behave friendly towards Mr. Howse, and party should they, as is too probable, meet with him in their war Excursions, had induced us, and the Canadians, to comply with the urgent Solicitations of the Indians to leave Men and goods at this settlement during the summer; and on this Acct. only could this place have been thought worth maintaining. The Muddy River Indian Chiefs have promised not to molest Mr. Howse on his return from the flat head country; but declared that, if they again met with a white man going to supply their enemies, they would not only plunder & kill him, but that they would make dry meat of his body...."

On May 27, 1811, James Bird wrote to Joseph Howse remarking:

> *"All that is necessary to be said, is, that if a trade to the flat head country should prove to be sufficiently valuable to justify us in continuing it, through the additional dangers and difficulties with which it must now be attended, I rely on your making use of every means in your power to accomplish so desirable an object... I confess that I have but little hope of our being able to recross the Mountain with advantage; I have however not with standing, left several extra men, to be at your disposal, if past success had induced you to form a different opinion...."*

But Howse, after returning safely to Edmonton House, considered it too dangerous to recross the Rocky Mountains. He returned apparently to Acton house in February of 1811 and *"... to Edmonton house by 30 July 1811, with thirty-six bundles of good furs, and convinced also that it would be folly to repeat the venture in the winter of 1811–12. The route of travel went through the lands of the Piegan Indians to those of the Flatheads, and a fierce war having broken out between the two, any white man who tried to traverse the country did so at his peril."*

During the summer a shortage of provisions developed at Edmonton, so that although Auld dutifully asked the committee if they wished Howse to resume his expedition to the Columbia, it was already out of the question for a year at least. Bird, at Edmonton, none the less went on planning the attack. He hoped to find a new route through the mountains, a route which begins *"near a principal source of the Athapascow River and terminates on the west side of the Mountains at the Cootakna River which, rising nearly west of the Head of the South Branch of the Saskatchewan, runs a very considerable distance in a northerly direction, makes a circuitous bend and turning southward joins the flat head and other rivers which together are supposed to form a very considerable part of the Columbia."*

Bird's plan was put into cold storage while Howse visited England in the winter of 1812–13. Here he was interviewed by the committee and was given a present or 150 GBP for his services *"in proceeding from York Factory to the Stony Mountains in the years 1810 and 1811... as a gratuity for his past and encouragement for his future exertions."*

According to J.B.Tyrrell, Howse had used a route followed by David Thompson in previous years. Howse ascended the North Saskatchewan, crossed over Howse Pass, and in the interests of his assignment, began his ascent of the Kootenay [Columbia] River, which he ascended to its head. From there, according to Tyrrell, he had gone to the Flathead River north of Flathead Lake *"not far from the site of the present town of Kalispell in Montana,"* where he wintered. Alexander Ross, when on his snake country expedition of 1824, halted on 14 February 1824, for his party to smoke *"at a spot on which some faint traces of civilization were to be seen."* This spot was on Riviere aux Marons, or Wild Horse River (now called Jocko River).

Ross added:

> *"A Mr. Howse, an entertaining individual belonging to the Hudson's Bay Company established himself here in 1810; but after passing part of the winter, he recrossed the mountains again, and never returned."*
>
> *I believe this is the first and only instance in which any of the servants of that company had penetrated so far to the west, prior to the country falling into their own hands in 1821.*

A map of the Columbia River by Alexander Ross dated Red

River Settlement, 1 August 1849 is in the collection of Additional Manuscripts in the British Museum. This map located, in spite of Ross's remarks quoted above, "Mr. Howse 1810" on the west shore of Flathead Lake, towards the northern end, and not on "Riviere aux Maron", which is also marked on the map.

Both Joseph Howse and William Hemmings Cook, the Governor of York Fort, apparently kept in touch with Joseph Colen, a former H.B. Company officer retired to Circencester. Letters from both men to Colen referring to the crossing of the Rockies in 1810–11 must have been sent to England by the ship which sailed from York in the autumn of 1811 for on 29 December 1811 Colen informed Sir Joseph Banks, President of the Royal Society, of the undertaking. Colen quoted Cook as saying that Howse had travelled *"across the Stoney Mountains and explored a Country that European feet dah never Trod..."* Colen also mentioned that Howse had not *"laid down his tractk"* as he was *"not provided with Astronomical Instruments."*

However, the following maps are listed in the Catalogue of Archives, which was brought up to date annually as records arrived from North America.

> No. 58[x] *"Rocky Mountains Sketch of showing the connection of the Athapescow, Saskatchewan & Missouri Rivers by J. Howes 1812"*

And...

> "No. 82, Columbia River Sketch of by Josh, Howes 1815".

A foot-note explains that the *x* against No. 58 means that the map was sent to Arrowsmith, the mapmaker. Like many others sent to the Arrowsmiths, this was presumably never returned. No 82 had also been missing from the H.B. Company archives for at least forty years.

When Howse finally returned to England in the autumn of 1815 Governor Robert sample of the Hudson's Bay Company wrote from York Factory on 11 September to Thomas Douglas, fifth Earl of Selkirk, and founder of the Red River Settlement;

> ".... *Mr. House who goes home in the Prince of Wales appears to be a plain sensible man and may perhaps be able to give Your Lordship some useful information. As everybody has been at Corinth so neither has everybody been across the Rocky Mountains....*"

Howse's journals or letters written during his stay west of the Rocky Mountains have not survived in the H.B. Company's archives. In the early part of 1843 when Sir George Simpson, the H.B. Company's overseas governor, was in London the Oregon boundary question was obviously much to the fore because he wrote to Joseph Howse and James McMillan requesting them to supply him with information about the early crossings of the Rocky Mountains. Copies of Simpson's letters have not been found in the H.B. Company archives so the exact questions asked are not known, but on 9 February 1843 Joseph Howse replied from Cirencester as follows:

> "*I hasten to answer your queries as follows:*
>
> First - *I crossed the Rocky Mounts, in the Summer & Autumn of 1810 by the North Branch of the Saskatchewan- ascended the Kootonay Columbia River- Carried into the Flat Bow Kootenay River- decended the most Southly Bight of it- crossed to Flathead River where we build. In Dec. with a couple of my men I accompanied the Flat-heads to the head branches of the Missouri- returned to our House- in Feb. 1811 and then to the* Saskatchewan - *the follg. Summer.*

Second - *None certainly I was myself the first in the H. Bay. Co's service.*

Third - *NW. Co. had already established themselves among the Kootoonays and had not preceeded so far as I was among the Flatheads.*
Fourth: *I am not aware of any.*

In rummaging over my H. B. papers which I have not done before for a great many years I have found the enclosed which are at least curious – I know not how they came into my possession – but they may throw some light on our subject.

I shall be most happy to give you any further information in my power.
Congratulating you most heartily on your safe return to old England..."

Note: When a French trapper caught a damaged beaver he took the hair off the hide and put it in bags. This hair was used to make high quality felt for hats. The quality of the hats is determined by the amount of thickness of beaver hair; the young beaver, the winter or muscovite beaver which comes from the winter hunting and there is the low grade beaver called 'castor-veule' which comes from summer hunting and is also used for making hats. Also there is 'castor-gras' which is the beaver skin that contracted an oiliness from handling or wear by the Indians. The trappers separated the hair in different bags in an effort to recuperate the hide or hair. Castor also signifies a hat manufactured with just beaver skin. While a demi-castor is one in which there is a combination of beaver skin and another type of material of fur. Today still a quality hat is called by the amount of thickness of beaver hair in the making. In our days it is marked by a number followed by an x and ordinarily the color is the natural one of the beaver hat—silver belly, tan and dark brown. In old times trappers separated the hair in different bags. Some of the Indian pelts were in pretty bad state and were recuperated this way.

Note: There is part of the Aaron Arrowsmith map of North America on which Governor George Simpson added the dates of his arrival at different places on the route from York Factory to Fort George, Columbia River, in 1824. That part of the map bearing the date and other particulars is missing, but the portion used by Simpson clearly shows that Aaron Arrowsmith marked *"Howse Ho."* at the north end of *"Saleesh or Flathead L."* However it is noted that *"Howse Ho."* is not on the reproduction of the Arrowsmith map of 1795 bearing additions to 1811, 1818, 1819, 1820 and 1824 which accompanies *Fur Trade and Empire*.

HISTORICAL
PART II

Historians, who attempt to peruse early day exploration are frequently confronted with the tiring and sometimes impossible task of determining a specific location because of the name changes in the topography of the land.

The name, Flathead, of particular interest here, may serve to illustrate as the name was applied to several lakes and streams in Montana; and was used perhaps so freely that any effort to identify a particular locale by this name became a most perplexing task.

The issue in the instance, is whether or not present day lake Pend d'Oreille (When the Northwest Co. sought trade in this area) was ever referred to as Flathead Lake. Both the Thompson and Arrowsmith maps identify Lake Pend d'Oreille as it was originally referred to by Thompson, as Kullyspellum Lake. Coues, however, is categorical in stating that present lake Pend d'Oreille, as well as present day Flathead Lake and the present Clark Fork River, were all referred to as ...... Flathead.

Coues states:

> *"Premising thus far regarding names, Thompson's movements in 1809–10 may be here outlined, in so far as they relate to the "Flathead"1. Of our text: "...here is where he* [Thompson] *built what he called the Saleesh house, sometimes Flat Head house, which we now see has nothing to do with our Salish or Flat Head r., or our Salish or Flat Head 1.,. or his Saleesh or Flat Head 1, but was on his Saleesh or Flat Head r., i. e. our Clark's Fork, in Montana."*

The name, Flathead, is found throughout the Alexander Henry journals and is often used by Thompson in his journals and narratives. Both the journals of Henry and Thompson note that the Flathead called themselves Salish. Thompson used *"Saleesh"* for the main part, and Henry *"Saleeish"*.

Lewis and Clark in their expedition to the Northwest used the term *"Eoot-last-schute"* in referring to the Flatheads. This name is peculiar to and recorded in the journals of Lewis and Clark, and of Patrick Gass for September 1805. Lewis and Clark seldom used the name, Flathead, but on one occasion of note referred to the Tushepaw-Flatheads when they were at Traveler's Rest Creek.

Partoll, in his discussion of the Flathead-Salish Indian name speculates that *"Eoot-last-schute"* was probably a phonetic misrepresentation of *"In-Schute-schu"*, a tribal term meaning red willow (dogwood), which grew in great abundance in the Bitterroot valley and served to identify the Flatheads. The name, Flathead, according to Partoll, as applied to the Flathead-Salish Indians, was in *"translation of their tribal sign, which meant natural head, [as opposed to the Clatsops and Chinooks] signifying that there was no implication of physical mutilation"*.

Howse, in his reply to a letter directed to him by Simpson in 1843, said in part: *"....crossed to Flat Head River ....."* But where was the Flathead R.? Arrowsmith did not give a name to the present Clark Fork River (above present Lake Pend d'Oreille), but placed the name, Flathead R., alongside the present Pend d'Oreille River which flows between present Lake Pend d'Oreille and the Columbia River. By contrast, Thompson on his map, labels the entire river system (the present Flathead, Clark Fork, and Pend d'Oreille Rivers) as the Saleesh River. Neither the Thompson nor the Arrowsmith maps gave a name to the rivers emptying into Saleesh Lake, the present Flathead L.

The Arrowsmith map unmistakably labeled present day Flathead Lake, as Saleesh or Flathead Lake and Thompson used, Saleesh Lake.

At the time Howse wrote his letter to Simpson in 1843 the main river system emptying into present day Flathead Lake was referred to as De Smet River. Apparently the rivers emptying into the present Flathead Lake did not receive their current names until about 1870 as the map of the Territory of Montana in 1865 and the DeLacy map of the same date show the *'Maple River'* north of Flathead Lake and *'Flathead River'* south of the lake.

The Clark Fork of the Columbia River named in honor of Captain Clark of the Lewis & Clark expedition, has had many local names, some of them still in use. The Clark Fork is formed in the Deer Lodge valley by the meeting of Silver Bow and Warm Springs Creeks and is known there as the Deer Lodge River. Though local parlance it then becomes the Missoula River for old times until it absorbs its big tributary, the Bitterroot River, after which it is finally accepted as the Clark Fork.

Thompson's map shows the present Clark Fork above its confluence with the Flathead River as a continuation of the Bitterroot River. This is the stream referred to by Thompson as the *"South Branch"*. The Clark Fork above its confluence with the Bitterroot River is designed "Courters River" and was considered a tributary of the *"South Branch"*.

Thompson's Saleesh or Flathead River, as he referred to it, began at the outlet of present Flathead Lake, was joined by the *"South Branch"* (Clark Fork), and from this point on to where it joined the Columbia was known as the Saleesh River.

In an effort to persue the travels of Howse west of the Rocky

Mountains a brief discussion of the habits and locale of the Indian tribes in the upper Columbia Drainage is necessary although in this manuscript no detailed analysis of Indian habits and culture will be attempted.

Malouf, in his manuscript, *Economy and land Use by the Indians of Western Montana, U.S.A.*, includes maps in his discussion to show the land domain of the Kutenai, Upper Pend d'Oreille (ear bobs) or Kalispells frequented the area along the present Clark Fork River and this tribe was closely allied with the Flatheads (Salish).

The Kutenai domain, large in extent, encompassed the area now known as the Flathead valley and roughly the northern half of Flathead Lake.

To date there are no calendar records available of the Kutenai tribe to confirm or disprove Howse's presence in the Flathead. Malouf and white's work in conjunction with Baptiste Mathias on the calendar records of the Kutenai go back to 1826 and any records or method the Kutenai may have had in preserving their history prior to this time have not been uncovered.

The main camps and routes of travel of the Flatheads, Kutenai, and Upper Pend d'Oreille, in and about their respective domains and over the mountains to the buffalo are of particular interest, as the Indian guides employed by the early traders, would undoubtedly follow the respective tribes most common route of travel. In the interests of this paper a listing of the travel routes most frequented by the Flathead (Salilsh) follows.

The following extracts from Malouf's *Economy and land use by the Indians of Western Montana, U.S.A.*, refer to the Flathead Indians travel routes leading from the Bitterroot Valley:

1. *"To Deer Lodge valley via Skalkaho pass."*
2. *"To the Big Hole country."*
3. *"Via Missoula up the Clark Fork."*
4. *"To follow up Pattee Canyon to join the Clark Fork at Bonner."*
5. *"To follow up Burnt Creek to the east to emerge on the Clark Fork at Bonner."*
6. *"As it is today, Garrison was a junction of two roads (the junction of two important river valleys). From here one could travel up to Deer Lodge valley along the Clark Fork, or they could travel up the Little Blackfoot and over the continental divide to Helena. From the Deer Lodge valley the Indians could go on to Butte for point between present Boulder and Butte, and emerge in the vicinity of Three Forks for hunting. The route through Helena also provided access to the Three Forks area."*
7. *"Trains to the upper South Fork of the Flathead River form the Jocko valley, near Arlee, and from the Flathead valley near St. Ignatius were routes to the white River country and beyond to the Missouri River drainage."*

The center of the Flathead Indian life was the Bitterroot valley, when travelling from the Bitterroot valley to the eastern area the Flatheads had a choice of routes. The selection depended for food and other advantages the particular route would afford.

PROBABLE ROUTE

Thompson, according to his journals, ascended the Fisher River in May of 1811 to what probably is now known as Mckillop Creek, and having followed up this stream, no doubt crossed over to what is now Loon Lake.

Howse may have followed much the same route as Thompson, ultimately to descend the course of the Thompson River. Other possible routes would include following up the course of the Fisher River and to follow one of its three tributaries; the East Fisher, the Silver Butte Fisher, or the Pleasant Valley Fisher.

Another route to be considered would be the Wolf Creek drainage with the possibility of passage over the divide to the headwaters of Ashley Creek, or into Pleasant valley; although if Howse chose to go eastward from the Fisher River, the route now followed by Hi-way #2 would be more in agreement with the Indian Road of his day.

If Howse wintered on the present Clark Fork River the chances are good that he departed from this area by going up the Clark Fork with the Flatheads (Salish) and followed one of the Indian Roads described on a preceding page. The departure route would be in agreement with Howse's letter to Simpson when he said: *"....I accompanied the Flat Heads to the head branches of Missouri ....."*

Henry's journal mentioned that McMillan and Montour had left Flathead Lake on Dec. 12th. Where they note *"the H. B. Co. are also settled"*. Howse also contended that he left his winter location in December to return to the eastern plains. However, neither the Howse or McMillan parties, despite the fact that they were returning to the eastern plains during the same month, mentioned the activities of their rivals with regards to the return journey.

The most eminent reporter of that day, David Thompson, had, in the course of his duties, gone to duties, gone to Rainy Lake House during the summer of 1810 and consequently did not have contact or immediate knowledge through the messenger service of the fur trading era as to the whereabouts of his H. B. contemporaries (Howse) to report in his journals. The other prominent journalist, Alexander Henry, made a detailed report of the departure of the Howse expedition, but there is no mention of Howse's return to Acton House as Henry was at this time on a strenuous journey of his own up the Saskatchewan and over Howse Pass into the Rocky Mountains.

Alexander Ross while on his Snake River country expedition of 1824 speculates or concludes as to the whereabouts of the Howse party in his narratives, not his journals, when he mention the "faint traces of civilization" at some point on the present day Jocko River.

Thompson, in his journals (May 1811), made note that he received information to the extent that the *"H. B. are in the lake"* but no further mention is made with regards to the Howse party. The brief notation, *"in the lake"*, would apparently refer to an area well within the acquaintance of the writer, and must have been in reference to the present day Lake Pend d'Oreille (Kullyspellum), as (Haywood's emphasis added) Thompson, on this date, has not yet seen present Flathead Lake. Thompson's notation together with the statement of McMillan in the Henry journals, would be substantial evidence that the Howse party was established on or above present Lake Pend d'Oreille, and not north of present Flathead Lake.

James Bird must have expected Howse to return in the same

manner or route that he had taken into the Columbia as Bird's 25 March journal entry of 1811 states that he sent two men to Acton House and from there to proceed to Cootanha River with horses and supplies for Howse.

Bird's Edmonton House journal entry of 17 Feb. 1811 reflects virtually the same news as reported by Henry:

> *"...two of the Nwt. Coy. Clerks Arrived at Acton House on the 23rd Jany. From the places at which Mr. Howse is wintering which they left on the 12th. Decr."*

Note the time consumed for the trip was 42 days.

Howse left his winter quarter sometime in December, as Howse reports he arrived back *"at our house in Feby. 1811, and returned to the Saskatchewan in July* [30] *of the same year."*

Bird, at Edmonton House, however, did not appear to be aware of Howse's return to Action House, however, for Bird reports approximately three months later from Edmonton House in his journal entry of 13 may 1811:

> *"Men arrive form Acton House. Wm Flett and four men remain there to pass the summer, a desire to conciliate the minds of the Indians, Mr. House, and party should they, as it too probable, meet with him in their war excursion"*

Howse in his letter to Simpson mentioned that only a couple of his men returned with him. He states:

> *"In Decr. With a couple of my men I accompanied the Flatheads to the head branches of the Missouri."*

Presumably the freeman with the Howse party, by virtue of their status, went their respective ways when their services were no longer required. But what happened to the other seven men listed on the H. B. Com payroll?

Bird's Edmonton journal may serve to explain:

> "Received letters from Mr. Howes, dated Cootana River 20th. August. 1810, in which he states that ...."

Howse may have attempted to correspond with Bird at Edmonton House and dispatched some of his men as messengers.

CONCLUSION

There remains but scant evidence from which to speculate upon the actual journey taken by Joseph Howse beyond the southernmost point of latitude on the Kootenai River. The maps Howse made in 1812 and 1815 and consigned to the map maker, Aaron Arrowsmith, were never returned to the Hudson's Bay Co. All of the Arrowsmith maps were sold at public auction on 28 July 1874 following the death of John Arrowsmith in May 1873. The Hudson's Bay Archives no longer retain any of the material concerning the actual journey of Joseph Howse and the personal papers of Howse were consigned, undoubtedly, to one of the five beneficiaries of his will and their whereabouts today are unknown.

Howse received little or no credit for his contributions to the Arrowsmith maps. Two of his maps are listed as being sent to the Governor and Committee in London.

There remains the report of McMillan and Montour form the Alexander Henry Journals upon their return to Rocky Mountain House wherein they state that: *"he H. B. Co. is also settled on*

Flat Head Lake."

If Howse wintered on the present Clark Fork River he most likely departed for the plains country via McDonald, Cadotte, or Rogers Pass, as Howse stated:

> *"I Accompanied the Flat Heads to the head branches of the Missouri... "*

According to Alexander Henry, McMillan and Howse were together at the old Kootanae House and each of the men certainly had some knowledge of the other's whereabouts between September and December of 1810. Strangely, however, when Howse and McMillan departed for the eastern plains, each chose to go by a different route; McMillan travelled back up the Kootenai R. and Howse travelled eastward to cross the Rockies.

Despite the fact that the Arrowsmith placed Howse House on their map at the head of present Flathead Lake, the evidence weighs more heavily in favor of Howse House having been established along the present Clark Fork river (above present Lake Pend d'Oreille). Howse's apparent relationship with the Flatheads is in keeping with the established domain of the Flathead Indians. The report of McMillan and Montour would point to a Clark Fork River location, as well as Thompson's reference to the H. B. *"in the lake"*. Howse's own report that he came out on the plains in the vicinity of the head-waters of the Missouri is in agreement with the long established route of the Flatheads (Salish) in their migration to obtain the buffalo, and the Flathead River, by definition, was river system between the mouth of Flathead Lake and the Columbia River.

From the information available, Howse was detained at the old Kootenay House by the belligerent Piegans and resumed his journey southwards sometimes in September. Allowing the expedition two weeks to arrive at their destination either present day Flathead or Lake Pend d'Oreille there would remain only about two months to establish a trading site. Should this premise be correct it is not likely that Howse House was a building of a permanent nature.

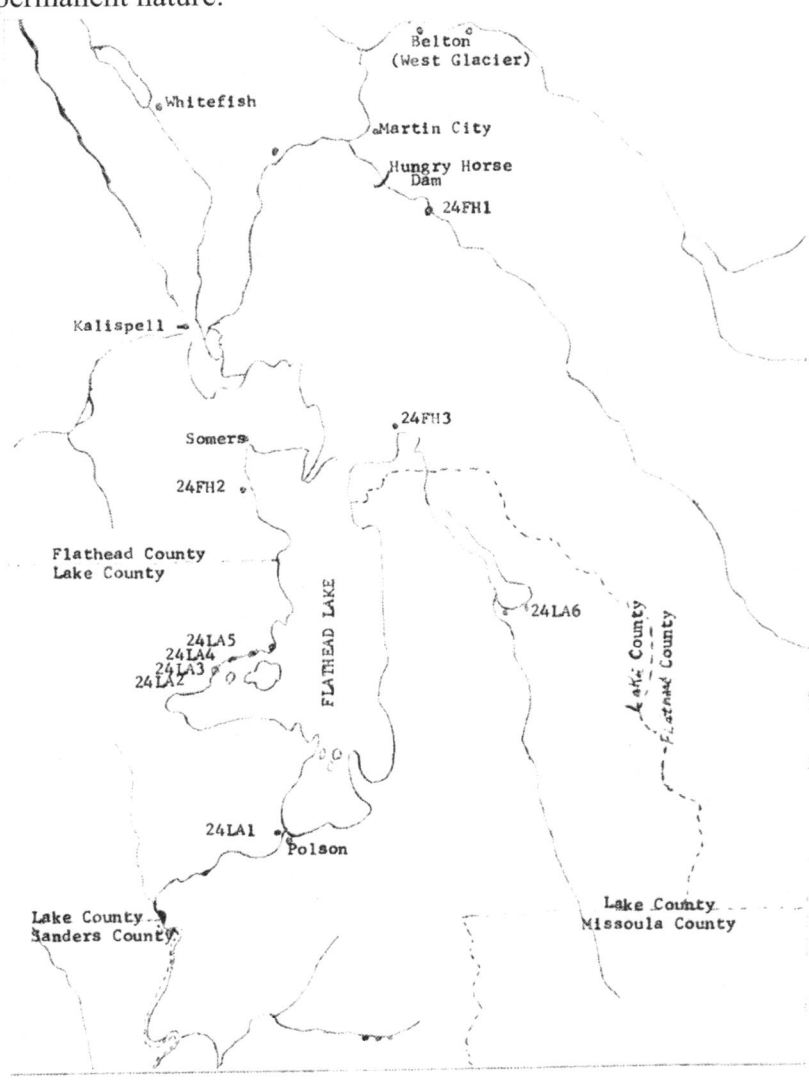

ARCHAEOLOGICAL SITE, FLATHEAD LAKE REGION, MONTANA
COMPILED UP TO 1953

FIELD SURVEYS BY AUTHORS IN
    FLATHEAD VALLEY

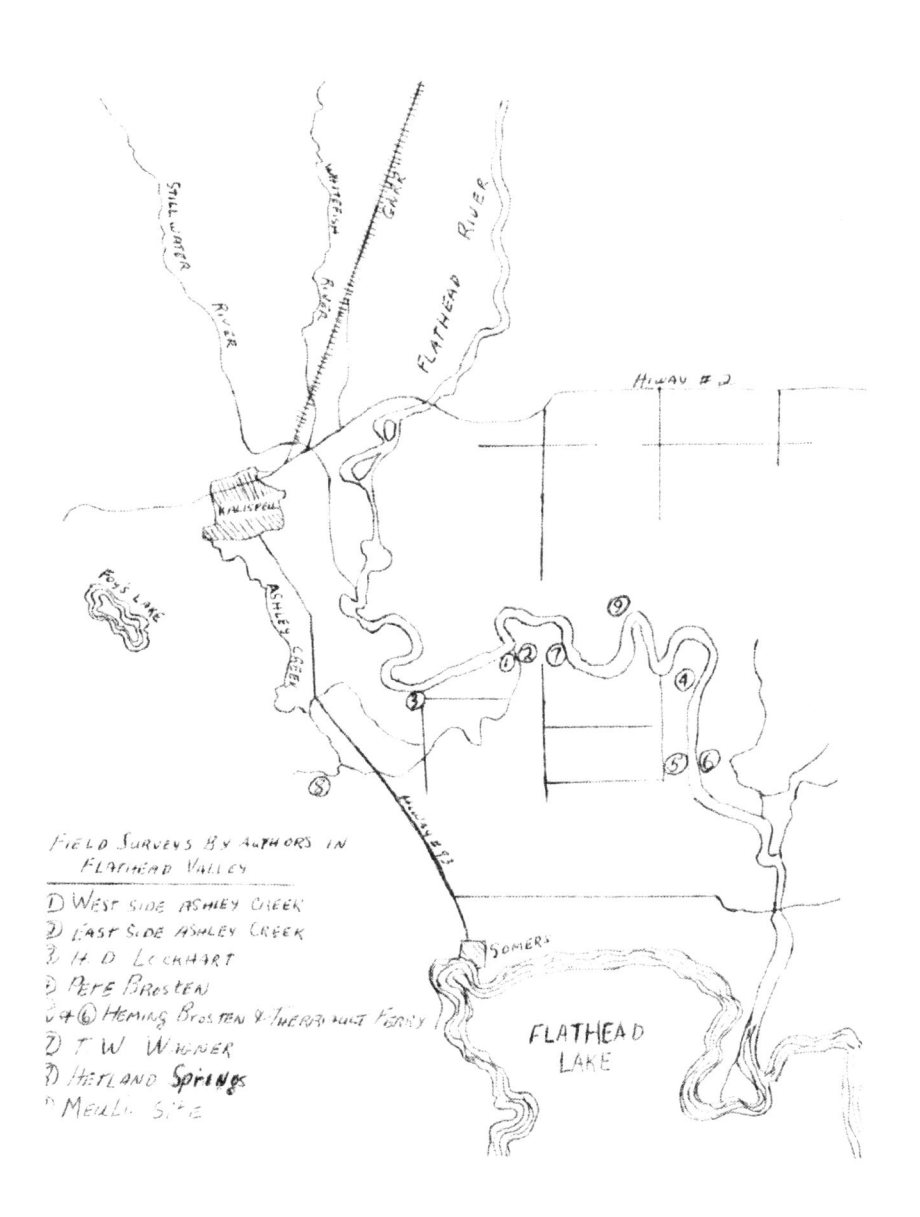

1) WEST SIDE ASHLEY CREEK
2) EAST SIDE ASHLEY CREEK
3) H D LOCKHART
4) PETE BROSTEN
5&6) HEMING BROSTEN & THERRIAULT FERRY
7) T W WAGNER
8) HETLAND Springs
9) MEULI SITE

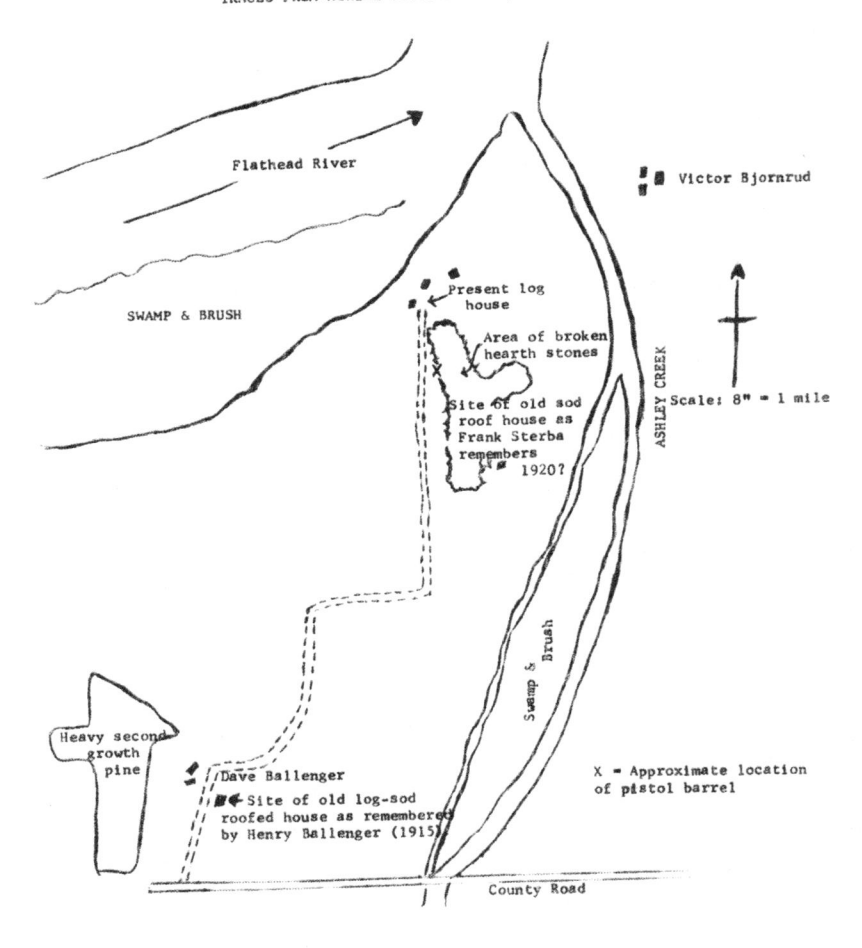

ARCHAELOGICAL SITES
PART III

There are a total of nine archaeological sites surveyed and numbered in the area around Flathead Lake. This includes a site now covered by the Hungry Horse reservoir, one on the Swan River and one at the head of Swan Lake. This includes all listings up to and including 1953. The Smithsonian Institute sent a group of trained archaeologists to Flathead County and they established a site near the present Hungry Horse dam. This survey by the Smithsonian Institute was conducted before the reservoir was filled, presumably in the 1940s. The number of the site is 24F.H.I. The number 24 designates Montana; the F.H. denotes Flathead County and the last number signifies the specific location. The survey number, 24FH3 is located on the Sudan farm and was reported by Miss Sudan several years ago. It is uncertain just how many sites have been surveyed if Flathead County by conscientious amateurs and reported to the proper authorities. There are, of course, the sites that are never reported and the 'loot' disturbed and scattered in such a manner that any evidence that did exist is beyond the reconstruction of the trained archeologist. The authors of this manuscript intend to list herewith their own personal observations in the field and to describe their findings in eight separate locations in what is known as the 'lower valley' (the area bordered by Kalispell, Montana to the north, Flathead lake to the south, by the mountains on the east and west of the valley.)

ASHLEY CREEK (WEST SIDE)

This site is located on the west side of the mouth of Ashley Creek and is situated in section 25, Township 28N, Range 21W. This site is located about six miles north of Flathead Lake.

The search for artifacts at this site produced a rather limited yield. A few badly scattered and broken heart stones were uncovered and a muzzle loading flintlock pistol barrel was plowed up by Mr. Henry Ballenger in 1955. The pistol barrel was located about 100 yards south of the present log cabin which is located immediately west of the mouth of Ashley Creek on the south bank of the Flathead River. The point of discovery was on one of the higher terraces that run parallel with Ashley creek in a north-south direction. The following map presents the general conditions existing there today. (1960)

MUZZLE LOADING PISTOL BARREL
FROM WESTSIDE OF ASHLEY CREEK

(Sketch is not to scale)

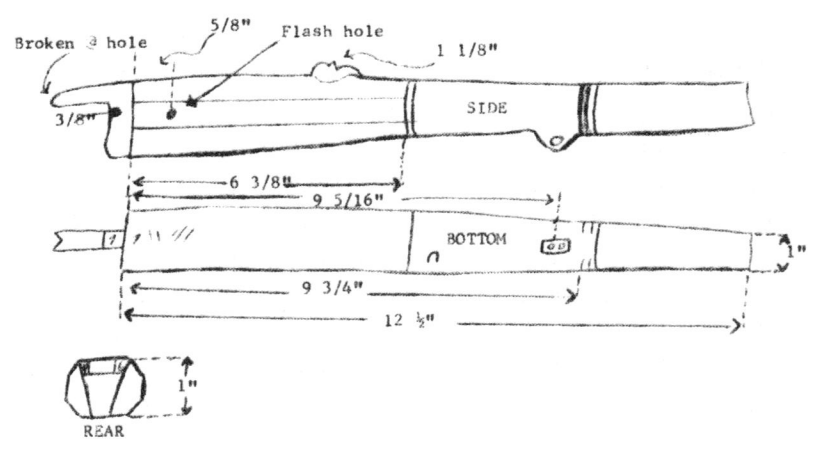

DESCRIPTION

```
Weight:  about 1 3/4 lbs.
Caliber: 0.562" or larger
A #4 is located on the left side of the barrel 1 3/8"
toward the muzzle on the 10 o'clock octagon. Also there
is a letter C on the bottom of the round portion of the
barrel 6 7/8" from the rear and a pair of matched 7s on
the plug and on the barrel.
```

H. D. LOCKHART SITE

This site is located on the H. D. Lockhart farm and is in Section 34, Township 28N, and Range 23W, and borders the south bank of the Flathead River. This site covers a rather large area. By estimate it is about a half-mile long by several hundred feet wide.

Our survey commenced near the Lockhart farm buildings and followed through in a semi-circle over into section 35 to the eastward. See traced map on proceeding page.

During the last 60 years there have been many lithic artifacts found on this farm land. Prior to the advent of farming this area was probably a light stand of pine and fir and was cleared in the 1890s or slightly later than this date.

The likelihood of this area being one of the possible location for 'Howse House' will be considered. First of all the Lockhart property has yielded many artifacts since farming was introduced to the area and it is quite possible that many artifacts have been found, lost, or not reported. Nevertheless, the Lockhart property represents a focal point of habitation established many years ago. For our purpose it could be considered as well within the area used as a winter camp by the Kutenai.

When a surface survey was conducted from the Lockhart farm buildings to the county road, many broken hearth stones, chips, flakes, and a few burned and broken bones were found along the higher ridges along with pebbles and rocks which are foreign to the cultivate land on this farm.

South of the concentration of hearth stones and thence eastward alone the ridges the survey revealed more lithic material. The concentration of lithic material seemed to lie in two places.

One was pole B, and the other about 46 yards south west of the north-south country road, and at right angles to the ridges or terraces. This particular site has revealed from previous years, 1 grooved hammer, 2 flat rocks, 1 net weight and several knives or projectile points. These artifacts have been collected by several parties over the past years and it is impossible to determine the full extent of the collections from this area. In addition to the previous mentioned artifacts our survey produced a few fragments of bone some of which were charred.

A good percentage of the broken hearthstones were found on the southern edges of the gentle sloping terraces south and east of the Lockhart farm. These hearth stones could possible indicate a winter camp facing towards the sun amid the openings in the timber on the heights of the terraces. This presumption would consider that the area to the north amid the slight swells and troughs contained considerable undergrowth or heavy timber, thus forming a natural shield to the cold north winds of winter. There are indications along these terraces of the remain of large trees as there are many fragments of heavy roots apparently of pine, tamarack, and fir. The northern edges of these terraces seem to contain many fragments of these roots.

To the west of the terraces, there is a concentration area yielding evidence of an old homestead or living quarters which is probably not old enough to associate any of its artifacts with "Howse House". The location of this "homestead" is perhaps significant as it was probably established in the 70s or 80s, and Howse House only proceeded these dated by fifty or sixty years. It is then entirely possible that the settler or squatter could have chosen for his home or quarters a spot of previous habitation if the area provided adequate space in the timber.

To date there are not any remains of structures such as piles of stones which could have been used for foundations. The survey, however, produced many pieces of window glass, crockery, broken bottles, broken pieces of horse farm machinery, and nails both square cut and round.

A good metal detector was used at the site of the old home and many pieces of metal were detected, including many unidentified portions of farm machinery. Also, there was an amazing absence or lack of tin cans either large or small. It is possible but not likely that the tin cans were removed to a dumping grounds; or the homestead existed before tin-can days, or it was entirely possible the survey did not locate any cans. There was not a fragment of a tin can to be found. Every piece of glass found was very highly iridescent and similar to the glass that was found at historic Fort Owen. The glass found at fort Owen was probably brought in before 1890.

In an effort to determine if there was any record of a homestead or house in this area a letter was directed to the U. S. Dept. of interior Bureau of Land Management, Billings, Montana. Their reply follows:

> *"No mention is made of any homestead or house or cabin in section 34, Township 28, Range 21W. I have looked this section up in the dockets section of the land office and there is only one entry for NE1/2, SE1/2 of section 34, Township 28N, range 21W. This particular subdivision was a cash entry and was patented to Ira B. Brun Feb. 1, 1893. According to the entry he had applied in 1891."*
> Signed;
> Mrs. V. A. Deeter,
> Clerk

ARTIFACTS FROM THE LOCKHART SITE

The Lockhart property has yielded many objects of archaeological interest beginning with the clearing of the land many years ago and the subsequent yearly cultivation to produce a rather sizeable collection of artifacts. These items will be listed and discussed.

1. A quartzite stone of measurements 3 x 4 x ½ inches. This is a natural water washed purple stone bearing three notches and appears to have been used as a fish net weight. Two of the notches oppose each other on opposite sides of the stone and appear to be of the same age, while the third notch appears to be more recent and no doubt is the result of the impact from some farm implement. The side of the stone which faced uppermost revealed a considerable deposit of calcium as well as one of the older notches. The more recent notch does not possess this deposit.

2. This stone, presumably a sinker, was found 60 yards west of the north-south country road near the top of the southernmost terrace among some broken hearthstones. At this location there appeared to be a concentration of lithic material, and there have been reports of the discovery of several of these atones or sinkers along these terraces by other parties.

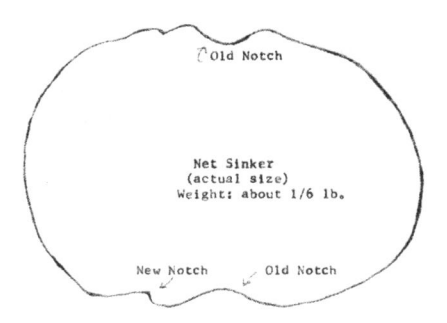

Net Sinker
(actual size)
Weight: about 1/6 lb.

Old Notch

New Notch   Old Notch

Across the Flathead River from the Lockhart property and to the north on the Siblerude farm about 35 of these sinkers were found in 1941; all of the sinkers were found in a pile when a new field was broken for farm use. This site is not described in this manuscript.

3. Two flat, water washed dorite 'cooking' stones, about one foot in diameter and one half inches thick were found at the same location of the Lockhart farm as was the 'net weight' in 1941 by Mr. Ernest White. A number of stones of the same shape were observed at a site on the Columbia River (east side) north east of Vantage, Washington by the authors in 1955. This is an archaeological site on the high levels of the Columbia River.

Mr. William Gingras, a younger Kutenai historian, viewed these two rocks before his death in 1955 or 1956 and his remarks follow:

> *"No, these stones are not for cooking. They are for the same as Whiteman's hot water bottle. You understand that among the early Kutenai, in older days, they designated a person within the camp to have such things. They are to be heated and used as a hot rock. You people call that a hot water bottle. You people call that person a nurse. The Kutenai had such people in many of their camps. Sometimes this person would go ahead of the main group when traveling and get these stones hot to be used on the person that might be sick, when that person arrived with the main group."*

4. A small grooved hammer of dimensions: 3 3/4" x 2 3/4" was found at the same location the cooking stones were located by Ernest White in 1941. This is a natural water washed

red quartzite stone and was found near the #2 concentration point. (See sketch of Lockhart site on page 103.)

5. Projectile points. These points, three in number, are simple medium sized (1 3/4" to 1 ¼" long) and stemmed barbed convexed base. They are about 1 to 1/2 inches wide and all are made of basalt. The points, slightly retouched on the forward 3/4 surface, were found by Ernest White, Harold Prouty, and Mrs. Ernest White on the Lockhart site somewhere between the power tine and the north south county road.

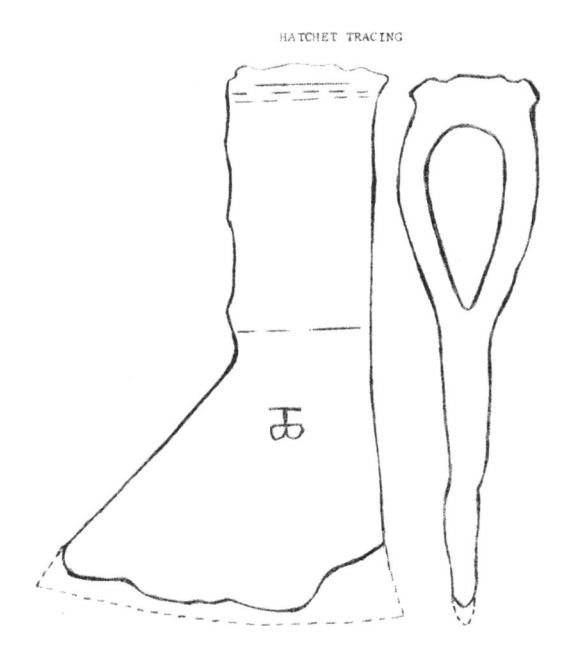

HATCHET TRACING

Long before this manuscript was considered (in 1950) a man brought the hatchet (traced above) to the Flathead Lake Lookout Museum statting: *"I found this over by Lockharts."* It is indeed interesting to note the stamp on it. Could it be Hudsons Bay? It is encased heavily with rust and clay and weights 1.5 pounds.

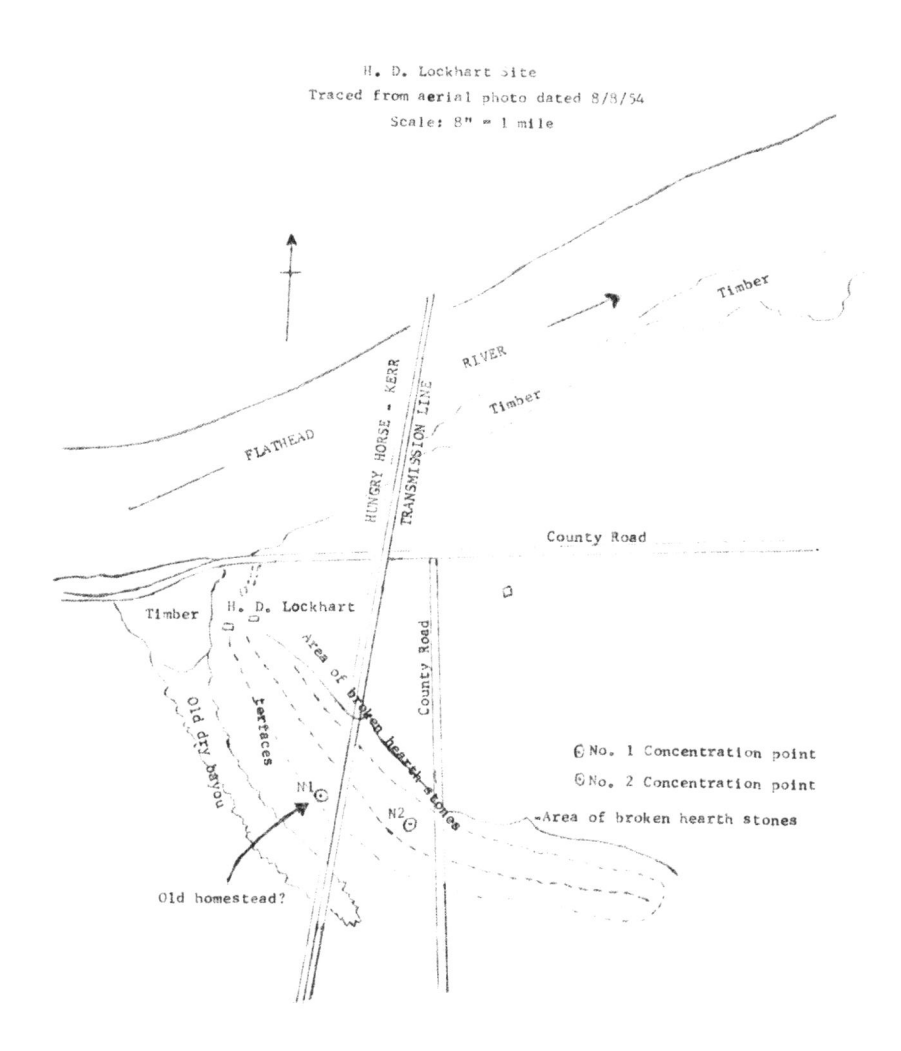

H. D. Lockhart site
Traced from aerial photo dated 8/8/54
Scale: 8" = 1 mile

EAST SIDE - ASHLEY CREEK SITE

This site retains considerable interest not only because of its potential in archaeological information but because of the historical significance of this area. A careful surface survey should be conducted in this area. To date nearly all of the area is under farm paractices. Our personal observations revealeed a few broken hearth stones.

EAST SIDE ASHLEY CREEK
TRACED FROM AERIAL PHOTO DATED 8/8/54

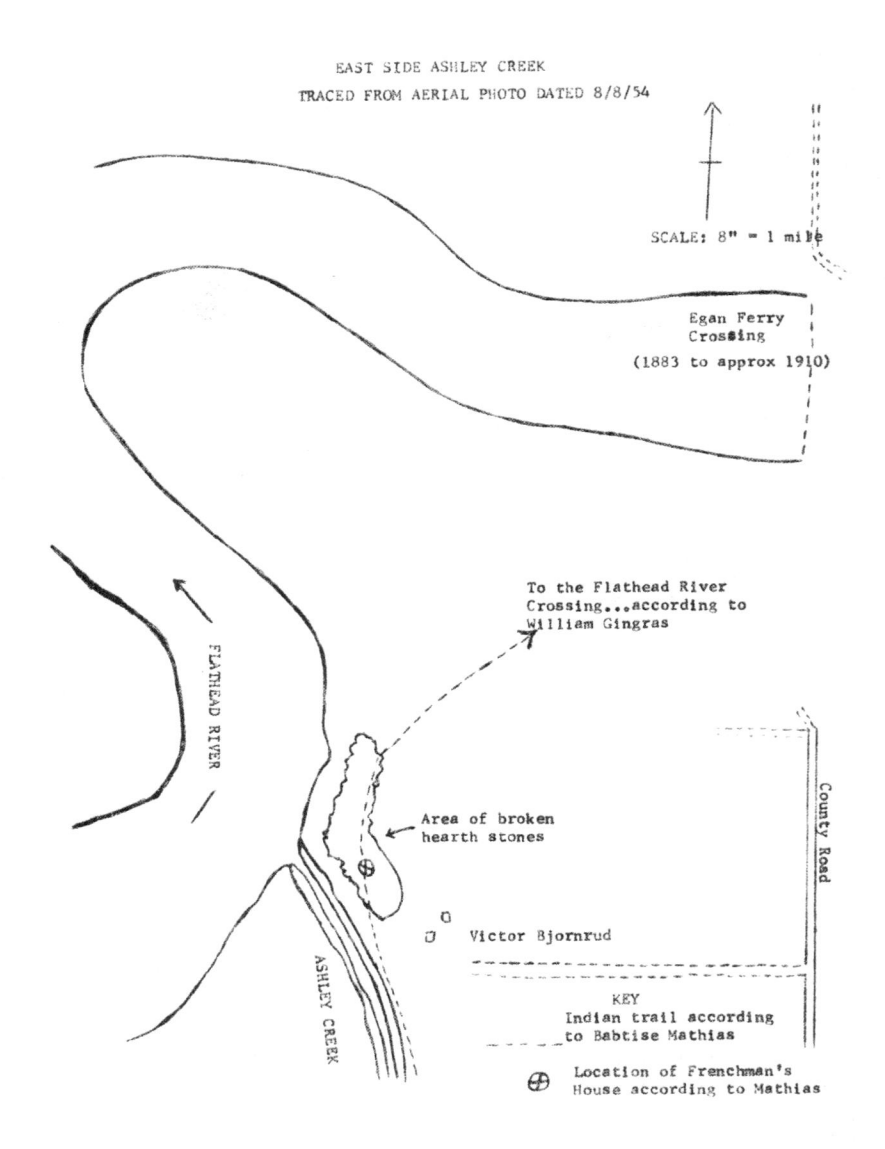

SCALE: 8" = 1 mile

Egan Ferry Crossing
(1883 to approx 1910)

FLATHEAD RIVER

To the Flathead River Crossing...according to William Gingras

Area of broken hearth stones

Victor Bjornrud

County Road

ASHLEY CREEK

KEY
— — — Indian trail according to Babtise Mathias
⊕ Location of Frenchman's House according to Mathias

104

PETE BROSTEN SITE

TRACED FROM AERIAL PHOTO DATED 8/8/54

SCALE: 8" - 1 mile

Area of new clearing

Section 28
Section 33

Concentration area

Old River Bed?

FLATHEAD RIVER

County Road

Pete Brosten

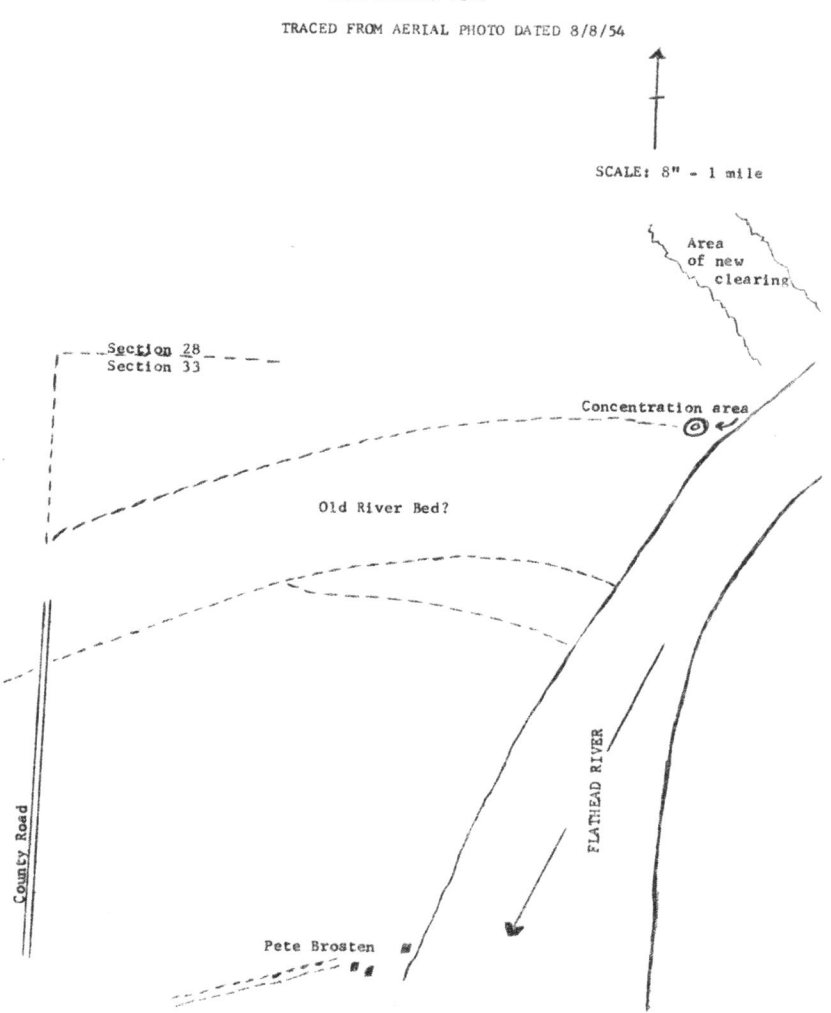

PETE BROSTEN SITE

This site is located along the west side of Flathead riverr in Section 33, Township 28N, Range 20W., and runs parallel with the river from Mr. Brosten's farm building up to and including a protion of Section 28. Nearly all of this area is being farmed, but some of the srea in Section 28 is just being cleared.

A simple surface survey was made from just north of the farm buildings up to the recent land clearing. During the past decade many lithic artifacts have been found on this site. One old pistol (its whereabouts unknown) was reported as having been located at the concentration point. At the concentration point numerous broken hearth stones were found, along with two or three simple water washed stones that were useful as pounders or smashers. Also there were flakes, chips and bone fragments at the concentration point.

Considerable time was spent in the newly cleared land but nothing of interest was observed. The concentration point is next to the river on a low terrace that runs east and west. This site warrants a more complete survey, as it appears to be the most recent of the sites described in this manuscript. The lithic industry aspects would testify to this effect.

SAND PIT SITE

The Sand Pit site is located on the north edge of the paved highway between Big Fork and Somers (Big Fork - Somers cut-off). Specifically, in the southeastern corner of Section 18, Township 27N, Range 20W. Considerable sand has been removed from this area for road construction and it is reported that several projectile points have been recovered from this semi-sand dune before it was hauled away. These items were reported to be surface finds. The authors have not visited this Area.

SUDAN FARM SITE #24FH3

This site is mentioned as it has been surveyed and assigned a number, and it is not too far removed from the Therriault Ferry site and is located just north of the northernmost meander of the Swan River above Swan Lake. To the best of our knowledge this area has not been surveyed with a metal detector.

For a complete description and list of artifacts located at this site see *Archaeological Sites in the Flathead Lake* p. 38-43 Anthropology & Sociology Papers #15, Montana State University, Missoula, Montana, G. Griswald (editor).

The Sudan farm site could very well be one of the main camps on the 'cross-trails'.

THERRIAULT FERRY (EAST SIDE) SITE

This site (see east side sketch page 108) is located close to the river bank and there does not appear to be a concentration point in this particular area. This site is located in Section 4 Township 27N, Range 20w. The only articles found at this site were a few scattered broken hearth stones.

HEMING BROSTEN SITE

This site consists of a scattering of broken hearth stones on the west side of the Flathead River, located just north of the old Therriault Ferry landing in Section 4, Township 27N, Range 20W. No concentration point was located at this site. However, several artifacts have been found north of the Brosten home along the river bank. In addition there are a few broken hearth stones along the river bank in the fields now being farmed. The artifacts as described were no doubt grooved hammers, arrowheads, war clubs, and the usual run of artifacts sought by arrowhead collectors. None of these items were available for inspection. A metal detector was used in Mr. Brosten's garden south of his home with negative results.

108

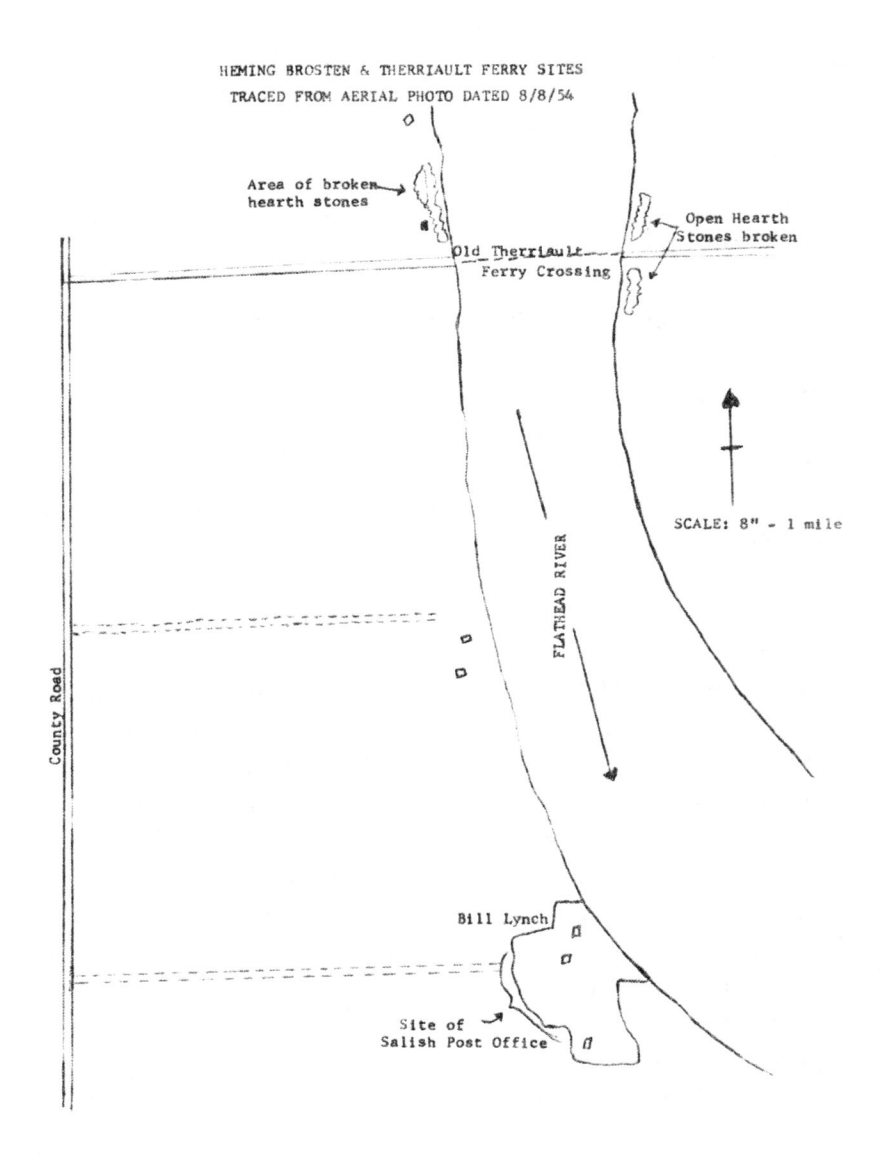

T. W. WAGNER SITE

In February of 1961 in the company of Mr. T. W. Wagner and his son, Robert, of Route #2 Kalispell, and Dr. Bruce Allison also of Kalispell, the remains of a former log structure were visited in the area proximate to the mouth of Ashley Creek. This site is located near the west bank of the Flathead River in the SE ¼, N. W. ¼, Section 30, Township 28N, and Range 20W.

Both Mr. Wagner and Mrs. M. E. Emmert (who lives immediately south of Mr. Wagner and who homesteaded this area with her husband in about 1900) tell of the remains of a cellar type structure, the walls of which were lined with logs placed horizontally to form a cache or container of approximately 15 x 10 feet dimension

The site as pointed out by Mr. Wagner is today nothing more than an indentation in the ground surrounded by an Impressive stand of birch trees, Mr. Wagner stated:

> *"When we started breaking the brushy ground to the west of this structure in 1932 we discovered, when we plowed, 40 or 50 places where there were many broken small rock piles. It looked as though the rocks had been placed in postholes. The size of the rocks probably averaged about the size of 2 or 3 inches. Every time we plowed these rocks would come to the surface. They looked as if they had been placed in same sort of pit or hole in the ground. We also found a broken grooved stone hammer near the structure (described above)."*

T. W. WAGNER SITE
TRACING FROM AERIAL PHOTO 8/8/54

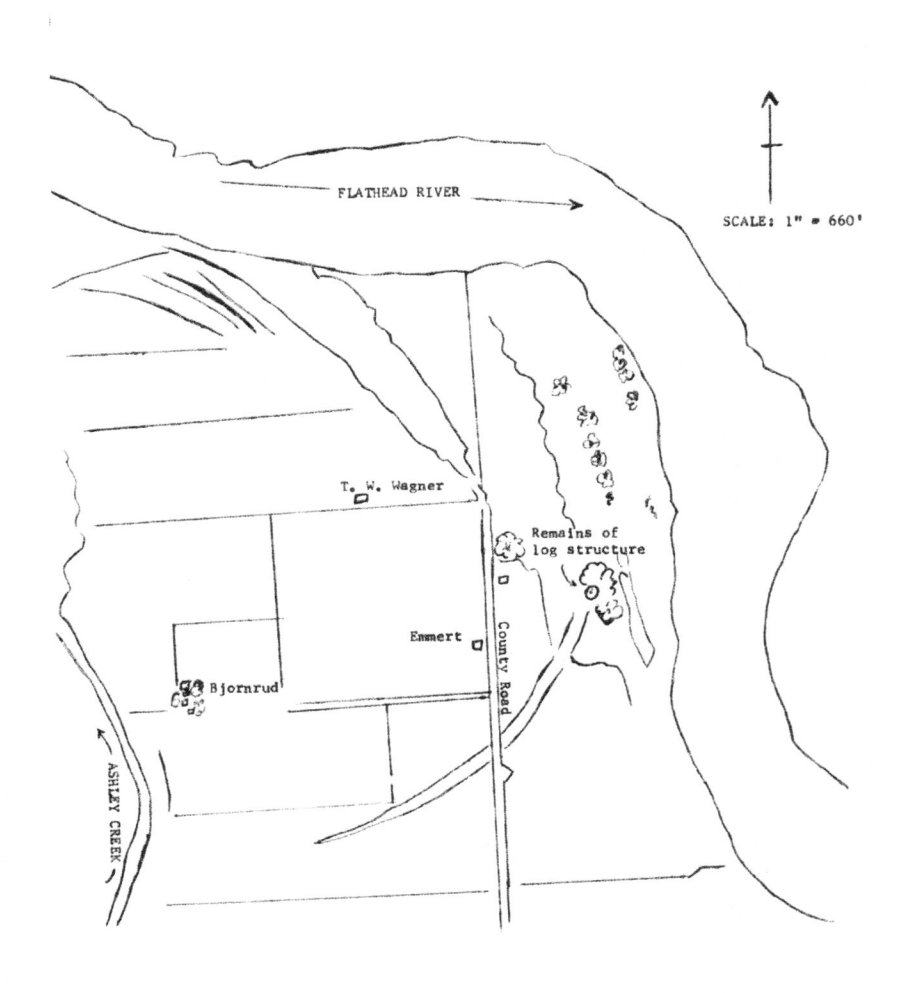

THE MEULI SITE

On May 21, 1963, a survey a conducted on a portion of untitled land at the southeast corner of the Selmer Meuli property (Sect. 35 T28N, R21W) by Thain White Dr. E. A. Allison and A. B. Braunberger. The central portion of the acreage surveyed contained to large mounds of earth each of which approximated 625 sq. ft. in dimension. These mounds of earth presented a curious and a most impressive sight as they were bordered by trenches approximately 25 ft. on a side and it was apparent that the trenches were at one time considerably deeper than they were on the data of the survey.

The trenches, mounds and surrounding areas were thoroughly surveyed with a metal detector and two 12 gauge Winchester, and two 16 gauge Remington shot gun casings were extracted from the soil together with a portion of a single bladed axe head. The soil cover was of sufficient depth to prevent an accurate appraisal of the metal content of the area.

Linear measurements were made of the trenches and recorded for the diagrams shown to accompany this discussion. Two sample cores were removed by an increment bore from a cottonwood tree growing from the west trench of Meuli Site #2.

Meuli Site #2 was located approximately 55 to 60 feet from the west bank of Ashley Creek and both Sites 1 and 2 were undoubtedly surrounded at one time by birch trees and the predominant chokecherry bushes which are most common to this areas.

Selmer Meuli on whose property the mounds of earth are located, said that the area contained the trenches and mounds when he was a boy (early 1900s) and that numerous rocks of tubular shape were removed from there some years ago and some of them were stored in his granary. Mauli contended that the area in question was referred to by old timers as The Trading Posts and to the best of his knowledge no one had made an intensive investigation of the mounds of earth. Further, that he was quite familiar with most of the farmland bordering Ashley Creek but he had no knowledge of any similar sites, along other lands adjacent to Ashley Creek.

THE HETLAND SPRING SITE

In 1962, the Hetland Springs Site was visited and a surface survey was conducted. Mr. Harry Hetland, on whose property this site lies, described and poited out the spot where the cabin was located

This site is on an open bench facing northward and is about thirty feet above the water level of a swampy area created by run-off from the Hetland Springs. The area investigated is noted by an "x" and circle of the tracing of the area from an aerial photograph (see following page).

This site produced the following items: 1 hand made (thumb portion) of a door latch which by estimate was made somewhere between 1890 and 1910, 1 doz. cartridges (1873–1900), 6 cartridges (1873–1879); some of these cartridges were Winhester, some Frankfort Arsenal, and some Remington.

This sitte as well as the open land across the swamp (to the north) appears to be one of the most likely spots for early day habitation in the Flathead valley.

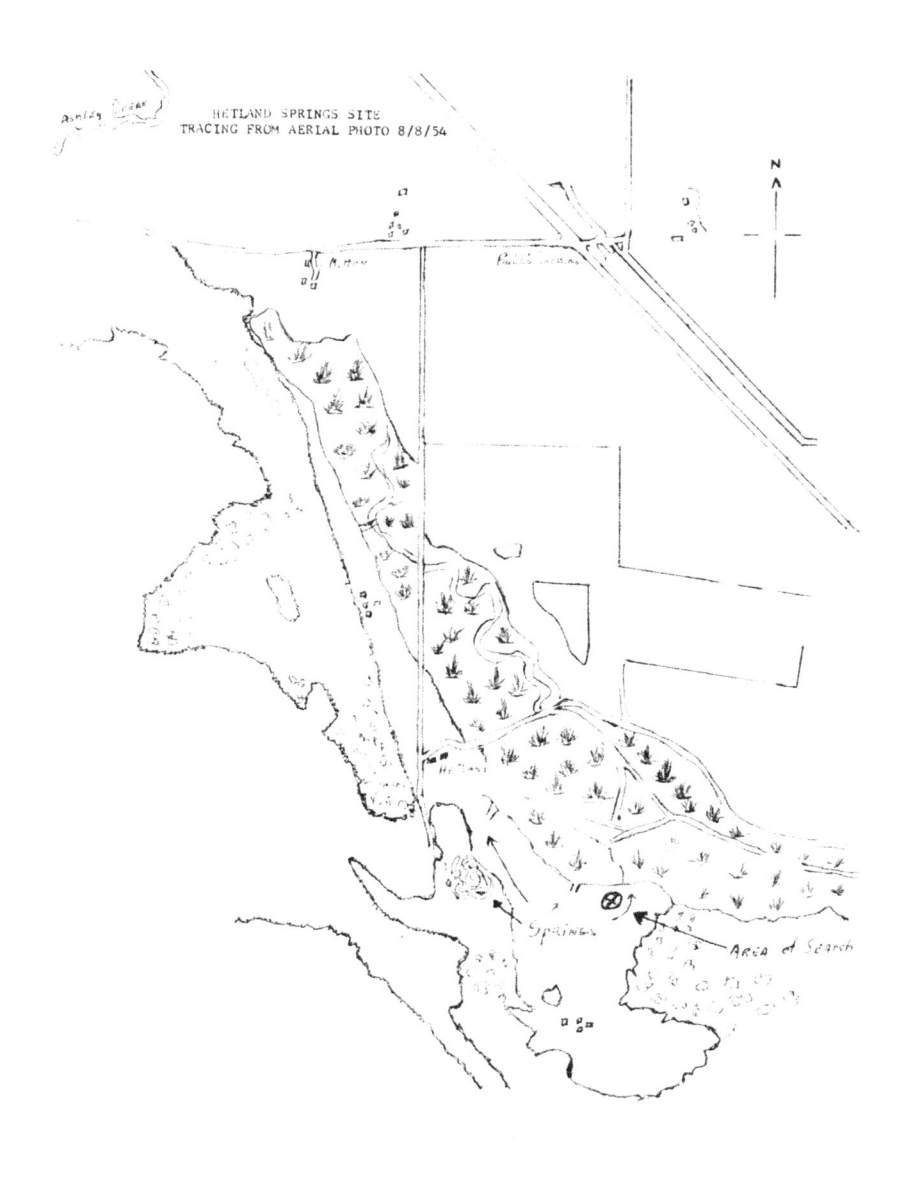

APPENDIX A

DERIVATION OF THE NAME KUTENAI

As the domain of the Kutenai tribe was centered mainly within the area known today as northwestern Montana, the derivation of the name, Kutenai, and the variations in its spelling will be considered in this appendix. The name, Kutenai, has been spelled, to the best of our knowledge, nine different ways:

> KOOTENAI COOTANIF OOTANAIE
> KOOTENAY COOTENAI KOOTANIE
> KOOTENAE COOTENEY KUTENAI

The appropriate, but perhaps offensive name, acquired by the Kutenai tribe is strange and unrelated to their own vocabulary. The name arose as a means of unmistakable recognition; to denote a physical characteristic common to this tribe and it was traced to the Blackfoot tongue.

The Kutenai name is a mocking and derisive term denoting... *"big stomach"*. The Flathead Lake band of Kutenai refer to themselves as the senka (sen-ka or sinka) tribe.

APPENDIX B

INDIAN 'ROADS' IN THE FLATHEAD LAKE REGION

The Indian trails (roads as they are usually called by the Kutenai) in this area apparently had been used for many generations. These roads were the lines of commuting (other than canoe) to and from their camping places, hunting grounds and many other localities pertaining to the economy and land use within their domain.

INDIAN TRAILS
FLATHEAD LAKE REGION

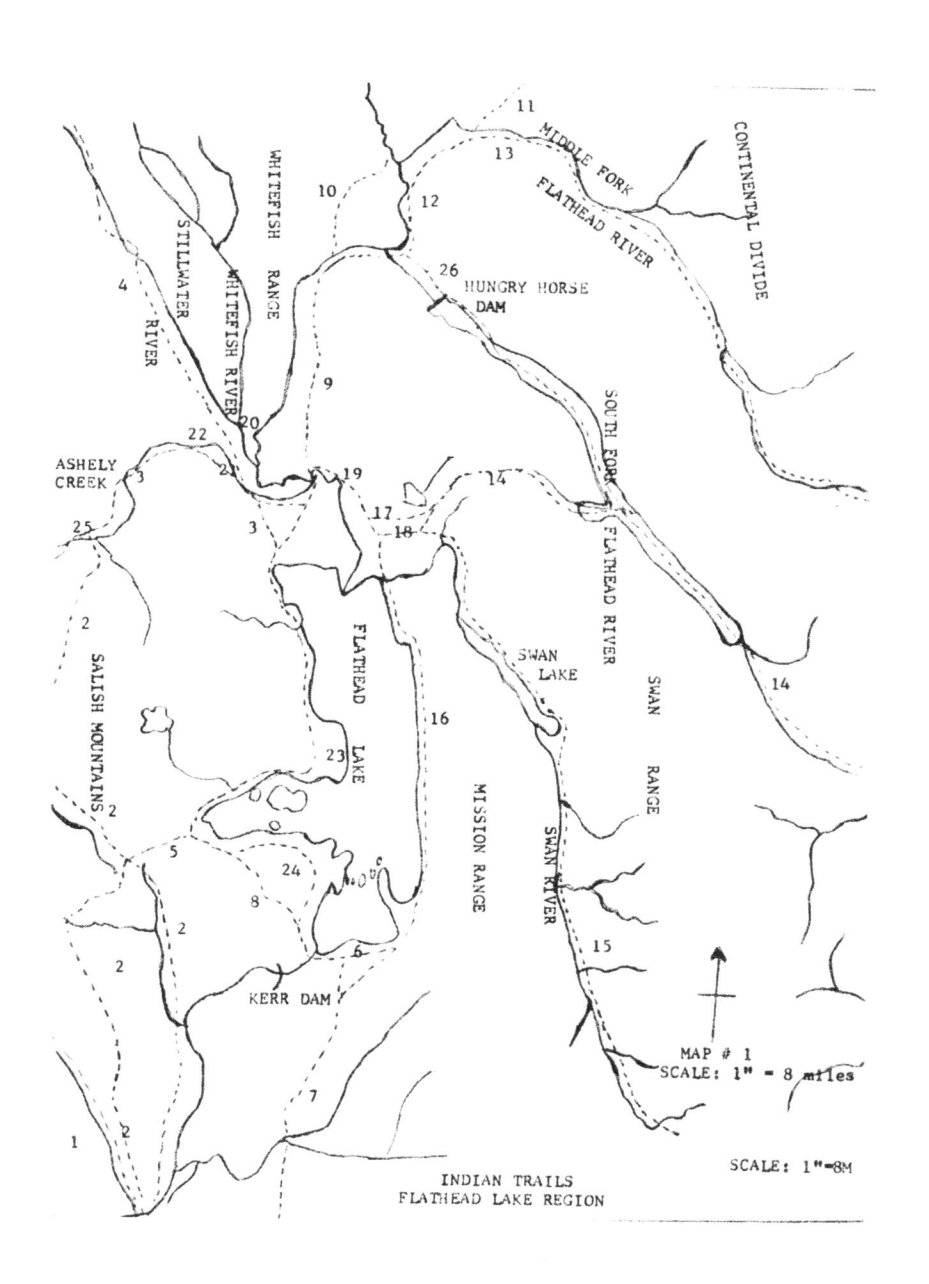

Even today some portions of these trails can be traced for several miles. Civilization had destroyed many of them. Many of the roads were used by the Kutenai up until about the time of World War 1.

To illustrate, these trails that are drawn out on the following pages are laid out based upon many sources of information too numerous to list. The information about the roads was gathered by Mr. William Gingras before his death in 1954 or 55. Mr. Gingras was compiling the names of camp sites and trails with the authors when his untimely death occurred. Some of this information he left with us is included here and portions of his maps are in the Flathead Lake Lookout Museum.

A list of Kutenai and mixed blood Indians is given here as our informant sources. Many interviews have been with these people by the authors since we have become interested in their history and folk lore.

Some of these Interviews as we might call them were perhaps our early acquaintanceship with them for nearly 30 years. Some over a longer period of time.

List of Indian Informants

Mr. William Gingras, Mr. Babtise Mathias (Last Sun Dance Chief of the Flathead Lake Band), Tony Mathias, Mitch Mathias, Mose Mathias, Lasso Stasso, Mrs. Left Hand (elder), Mr. Auld, Mr. Josephine Martin, Mrs. William Gingras, Mr. Babtise Left Hand, Mr. Basil Left Hand, Mr. Jerome (Nickolas) Hawanahorn, Mr. Abraham Bull Robe, Mr. Angus McDonald (Young Angus), Mrs. John Cacheure, Mr. Alex Cacheure, Mr. and Mrs. Joe Skookum, Mr. and Mrs. Charles Andrews, Mr. and Mrs. Joe Antiest, Mr. Walt Reidle, Jr., Mr. Wlater McDonald and several others whose names have slipped our memory.

Many other people that Mr. William Gingras had told us about while he was compiling material for his manuscript on The History of the Kutenai. It is unknown where this unfinished manuscript is now. It certainly contained an amazing amount of Kutenai history. Unfortunately, the manuscript was not finished before Mr. Gingras died.

Noting from conversations and interviews with local Indians the regions of departure from their winter camping ground; (around the head of Flathead Lake) many of the trails they consistently used seems to originate from very near the mounth of Ashley Creek.

    Indian Trails and the explanation given to us by
    Mr. William Gingras:

1. The road up and down the Little Bitteroot River from Perma to the Thompson's Lakes region and north down the Fisher River to Libby and Jennings.

2. The trail from Perma up Camas Valley to Hot Springs to Niarada then to Hog Heaven Hill, over the old cattle and sheep trail on Hog Heaven Hill to the head of Mount Creek on Browns Meadow then to Kila. — 2A. The trail up the Flathead River to White Earth Creek then to the Big Draw near upper Battle Butte School.

3. A trail from Kila to the old Ashley Creek Store (the site of Ashley Creek Store is nearly within the city limits of Kalispell).

4. The trail from Pot Holes to Tobacco Plains.

5. The trail from Elmo through the Big Draw to a place where Niarada is now.

6. A short trail across the foot of Flathead Lake.

7. The trail from the Jocko River through the Mission (St. Ignatius) and to Fort Connah to the foot of Flathead Lake.
8. A trail between Polson and Elmo going over Whiskey Gap.
9. The trail from Flathead River Crossing just north of the Winter Camps (mouth of Ashley Creek) to near Martin City on the South fork of the Flathead River.
10. The trail from Martin City to Canada via the north fork of the Flathead River to mountain passes for buffalo, Gingras stated: *'I have heard the older people talk about these but I do not know their names that you white people call those passes.'*
11. The trail from near West Glacier (known as Belton before Gingras died) up McDonald Creek to Logan Pass, this trail was used when going after buffalo. *'I have heard the older Kutenai talk about that'*, said Gingras.
12. A trail from near West Glacier (Belton) to Bad Rock Canyon on the Flathead River.
13. The trail that ran from West Glacier (Belton) and followed up thenMiddle Fork of the Flathead River to Maries Pass.
14. A trail from near Echo Lake that goes aver Aeneas Pass to Birch Lake then down to the South Fork of the Flathead River, to the Tobacco Plantations at Spotted Bear, then up the Spotted Bear River.
15. The trail from near Echo Lake to Swan Lake then on south to Cadotte Pass.
16. A trail on the east lake shore of Flathead Lake.
17. The shortcut trail across to Aeneas trail.
18. Another cross trail when going to Swan Lake and up the Swan River.
19. A trail from Flathead River crossing when going east, e.g. when traveling southward down the east lake shore of

Flathead Lake.

20. The same as number 4.
21. A short trail from Ashley Creek to the Tobacco plains Road.
22. A shorter trail from Ashley Creek to the Tobacco Plains Road.
23. The trail on the west lake shore of Flathead Lake. Between Elmo or Dayton, Mont and the head of the Lake, Medicine Rock, or 'Skinkoots' is on this trail.
24. The lower level trail going over Killawatt Hill used more than Whiskey Gap, According to Gingras: "You know the place I told you about where the Kutenai man froze to death a long time ago." The reference to the frozen Kutenai Man by Gingras is used to illustrate or point out a location on the lower level trail, a point near the top of the hill north west of Polson, Montana where U.S. Hiway # 93 starts on the level before going down bill towards Big Arm, Montana.
25. The same trail as number 3 going to the Fisher River country and the same country as trail number 1 on to the Tobacco Plains country.
26. The trail from near Hartin City to near Aeneas Creek on the south Fork of the Flathead River. This trail went up the north side of the river most of the way, and was used when there was too much snow on the Aeneas trail. Nearly all of this trail is covered by water as the result of Hungry Horse dam.

The intersection of these trails would be the logical areas to investigate with regards to early day habitation, as the Indian roads undoubtedly date bace into prehistoric times, even before the horse entered into this area.

According to the map (page 115) there would appear to be four or five trail intersection sites. Mainly the Flathead River crossing, the Sudan Farm site, the mouth of Ashley Creek, the old Ashley store location, and near the mouth of Swan River and slightly west of this point.

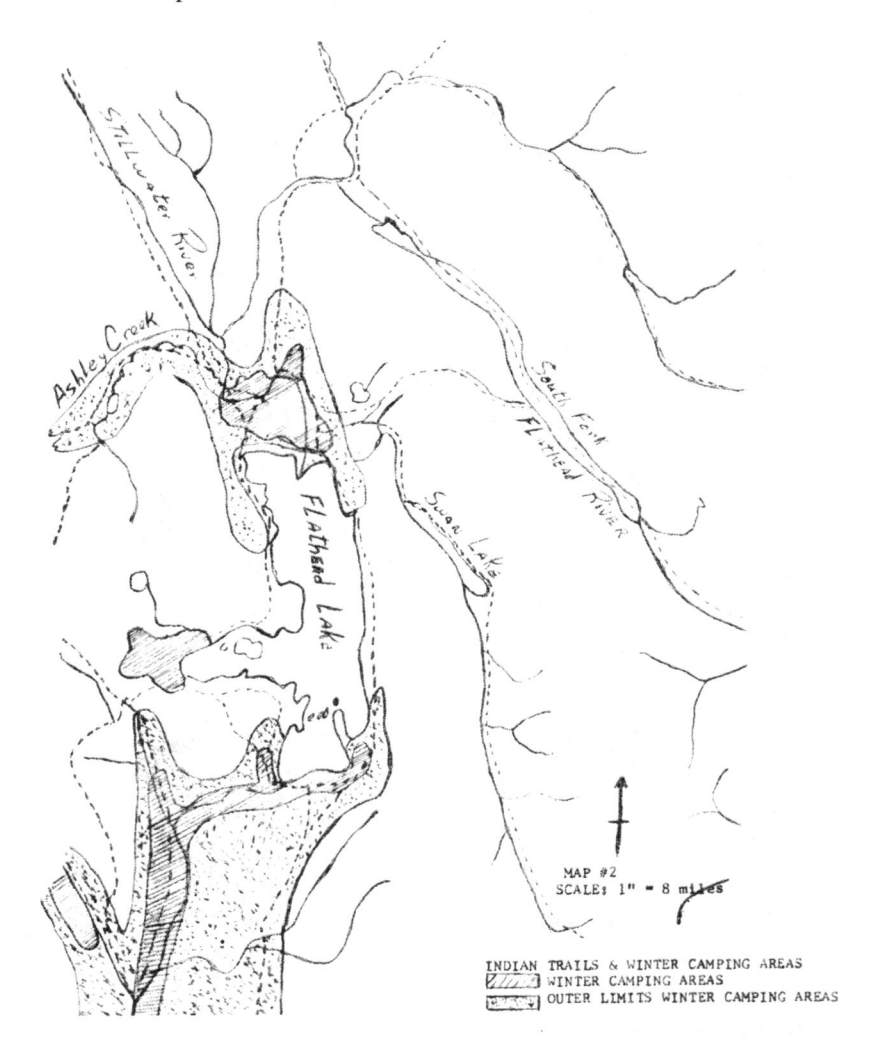

APPENDIX C

EARLY DAY LAKE AND RIVER TRANSPORTATION

In the preparation of this paper, the authors had occasion to observe the remarks of several historians regarding the dimensions of streams and bodies of water within the area concerned in this discussion. Mainly the statements were concerned with the natives' mode of travel in traversing the Flathead River and the possibility of crossing portions of Flathead Lake.

It was apparent from the statements that the myth or concept prevails, and perhaps justifiably, that the Indians east of the Mississippi River used the canoe mainly as a means of travel and the western Indians transported themselves over water, their families, and their baggage, only of necessity; to get from one point to another.

However, and in contrast to the above statement, it is evident that the bark canoe was used throughout the Canadian plateau; also in this country by the Colville, Spokane, Kalispell, Couer D' Alene and the Western Flathead tribes. These Indian tribes used two distinctive canoes; the sturgeon (nose type) and one designed with an undercut bow and stern. There was also a third canoe used by these tribes with the bow and stern lines almost vertical.

The Kutenai and Flathead tribes used the Plains Bullboat. The Klikitat tribe placed hides over a frame of uncertain form. The former bullboat was constructed by placing or lacing parfleches (rawhide envelopes) around the paraphernalia to be transported. On occasions brush was added for buoyancy. The Flathead bullboat was usually pulled by a horse, but among the Kutenai a swimmer tows the burden.

Ferris sneaks of the Pend d'Oreille Indians that he encountered on the banks of the Clark Fork:

> *"In a few moments the lodges disappeared, and the bosom of the river was studded with hark canoes conveying whole families and their baggage down the stream with surprising velocity. I was 'greatly deceived in their canoes, for the squaws would lift them from the water on to the bank, and again Set them into it with such ease that I imagined they must he quite insufficient to the transportation of any heavy burden. Some of them however, appeared loaded until there was no longer room for anything more, and still floated securely. They were managed by the sqaws, who, with paddles, direct their course with great steadiness, astonishing rapidity, and apparent ease and dexterity."*

APPENDIX D

DATA ON SCARRED TREES - ASHLEY CREEK AREA

Throughout the spring of the year when the sap of the yellow pine was commencing to run the Indians found the cambium or growing layer of the tree to their liking and utilized this stratified growing layer as a means of nourishment and a tonic as well, The abrasive process of removing this commodious layer together with the protective bark usually left a deep and characteristic scar upon the trunk of the tree, On occasions the Indians completely girdled the trunk and caused an untimely demise of the tree.

The valley floor in the immediate vicinity of the mouth of Ashley Creek yields sufficient evidence today that this area, before the coming of the white man, was abundant with the stately growth of the ponderosa pine. However, this majestic tree reluctantly gave way to progress; the axe, saw, and agricultural machinery, With the exception of an occasional lonely pine tree to be found along a slough or river bank, there are few stands or groves of ponderosa pine remaining. The surviving yellow pine trees; those untouched, and those with scars upon them, are today old and mature trees well over 150 years old.

An increment-borer was used in this study to determine the date the trees were peeled by the Indians for food, the increment-borer is an instrument used to remove a small core form the tree, and is fashioned quite similar to the coring machine as used in a diamond drill. The small core obtained from the tree is removed from the device and the rings are counted.

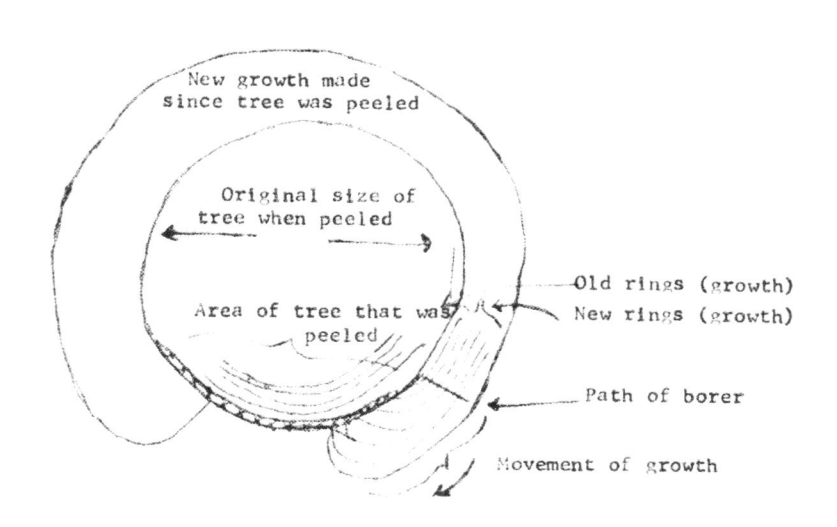

DATA ON SCARRED TREES – ASHLEY CREEK AREA

Tree Numer	Date Peeled
1	1868
2	1917*
3	1896
4	1837
5	1877
6**	1889
7**	1848
8	1828
9	1857
10	1857
11	1851

* This date is uncertain as the junction of the old scar and the new growth was difficult to determine.
**Same tree, two scars.

The age of one of the younger trees not scarred by the Indians was 142 years. In the yellow pine grove on the well property it is evident that many trees have been cut and removed; many of these trees may have yielded evidence of peelings as practiced by the Indians.

The Well farm indicates that this area was a rich source of cambium layer food; possibly as far back as 1800. However, the oldest scar revealed by this survey would appear to have been Cut during the year 1828; just 18 years following the visit of Howse to Montana.

APPENDIX E

KUTENAI INFORMANT SITES

The Indian Informant sites are included in this manuscript because of their historic importance, archeological significance, and their relation to the intent of this paper. The Kutenai historian and patriarch, Baptise Mathias, because of his knowledge of the early day Kutenai camping grounds, was convinced by Thain White that he should journey 'once again to the upper flathead valley and re-visit and most certainly point out the Kutenai camping sites.

This trip was made in the summer of 1959 and the first stop on the journey northward was a point slightly north and east of the present town of Somers, Montana. Here Matthias exclaimed:

> "Baptise Lapoo lived here near water...1884 or 1885. Him first white man me see. Maybe me five or six log house."

Continuing northward, the next stop was approximately half way between the present towns of Kalispell and Somers, Montana on Highway #93 and thence about a mile westward on a road leading to the foothill road, Here Mathias again reminisced:

> "Joe Ashley... he name for creek...maybe his daddy. I now know. Log house over hill from big spring. I there long time ago. Maybe me eight years old. Him good man."

At this point Mathias and white engaged in a lengthy discussion in sign language and Pidgin English and ultimately it was concluded that the approximate site of the Ashley cabin was near or on the site where an old log cabin, the remains all but rotted away, ware evident in the early 1930s. This site is about t mile west of the old Patrick Creek dam (used for saw mill), and about 3/4 mile south of the present Kessler ranch buildings. This site is located in Section 5, SW ¼, SE ¼, T27N, R21W. Again, and apparently referring to Joe Ashley, Mathias said:

> "Him first white man me secum, that's all me know about that place. I am tired,"

Frequently, and because of Mathias' age and the difficulty in communicating with him, it was most difficult to ascertain with any degree of certainty the details that were most desired; especially with regards to the site of the Ashley cabin and the derivation of the name—Ashley Creek.

Following this conversation Mathias was driven eastward to the area immediately bordering the mouth of Ashley Creek, The Ashley Creek site, according to Mathias, is the place where the two Frenchmen had a house and danced, The aged Mathias, apparently to make certain of his recollections, or because of diminishing vision, inspected the area to the east of the mouth of Ashley Creek intently. Then, while walking back and forth, he proceeded in the downstream direction along the Flathead River for a distance of about 300 yards, Mathias stopped and said:

> "Lotsa brush."

Mathias said no more until turning back and was about half way between the mouth of Ashley Creek and the Bjornrud farm buildings to the southeast, he remarked: *"Here creek."* and pointed in a westerly direction in a south to north gesture, Mathias continued. *"Old Indian road."*

Then pointing to the ground he said:

> "Come here good place to dance. You know two Frenchmen come. Have house, dance lots, yes? You know him? I showed you. Four Frenchmen come. Two years. Yes, you know him?"

Mathias and White continued their discussion in sign language and Pidgin English. Mathias digressing for the moment to discuss and make certain that his calendar records would serve to explain much of this discussion.

Mathias continued:

> "Have same house, same place. My daddy showed me. My daddy told me. Maybe me eight, maybe six. Lotsa brush," said Mathias pointing along the creek, "Open here, him good. Me no seeum house, all gone. My daddy says here. Hoy yea! Soyepie all over now. Kutenai all gone. Him land belongs to Kutenai. Me too old, tired, Taxas."

Mathias was questioned with regards to a store or trading post in this area and after lookin back across the river he stated:

> "Boat came her to us. Hi come from over there. Lotsa chokecherries here. Now all farm."

Mathias was then asked as to the placement of the Kutenai winter camps. He replied:

> "You seeum all round here. Good place for winter. Long time Kutenai use these places. Him good – Taxas."

In his talks with Mathias during the past decade White has

made many pointed inquiries as to trading posts in Northwest Montana. In particular, of course, to Howse House; and Mathias has always referred to Fort Connah or to the Catholic Mission at St. Ignatius. Neither Mathias or his calendar records reflect any knowledge as to a Hudson Bay Post north of Flathead Lake.

Present day evidence would tend to support the Ashley Creek area as being a favorable wintering area for the Indians, The prevailing winds from the south apparently are warmed in their passage over Flathead Lake and tend to favor this area with a more temperate winter climate than the adjacent areas.

Remaining in this area today are many old yellow pine trees with scars upon them that the Indians made when removing the cambium layer for food. This practice has been discontinued, but the growth of each year can be counted to determine the exact date when the bark was removed.

This practice of removing the cambium layer was conducted in historic times only In May and June. To confuse the issue there is evidence of trees being scarred in the spring at the winter camps which tend to confirm the speculation that some of the tribe did not accompany the migratory movements to other areas.

KUTENAI WORDS USED BY BAPTISE NATIVES

These words were noted and recorded in the fields during the trip described on the preceding pages.

Ka-poke-kan'-an-ni took – This Kutenai word refers to the winter Camping ground at the mouth of Ashley Crock, and no doubt refers to all of the area along the south side of the Flathead River from the old town of Demersville to near Holt.

Akom'- non-ok – Refers to a camping place on the northwest shore of Flathead Lake in Lot 4, S26, T27N, R21W.

Kookp'-Kousal-ouhs – This term refers to champ ground where the only water was avilable before arriving at Forey Creek (south bound), this camp was located at the south end of Glacier Bay on the west Shore of Flathead Lake.

Yak 'at-oul-ka – Refers to the camp at the mouth of Stoner Creek (Big Creek). This point is also referred to as Archaeological site. The authors compare their spelling of the Kutenai words used with those In Kutenai Tales.

Soya'pe – White man. This spelling is correct according to Boas, op. cit.

Nu la' qana – Frenchmen, This is nearly correct, however, our letter, L, Is too harsh and our third A is too broad.

Nu la'qana – Boas, op, cit., p. 363.

APPENDIX F

In the Booklet *(Historic Adventure Lend of the Northwest)* by Grace Flandrau put out by the Great Northern Railway there is a folding map title historic Adventure Land of the Northwest drawn by W. H. Gordenier 1926 and Copyright by the Great Northern Railway 1927.

The map is about 18" x 7.5" and shows the routes of early fur traders, the Great Northern Railway portions of the Great Lakes, Canada, California, Nevada, Nebraska and many, many locations of early day trading posts, forts, and major cities. The map is well executed and many posts, etc. are very accurately established considering the scale it covers.

Howse House is one of the points listed and its location is on the east side of Flathead River about where present day Holt is.

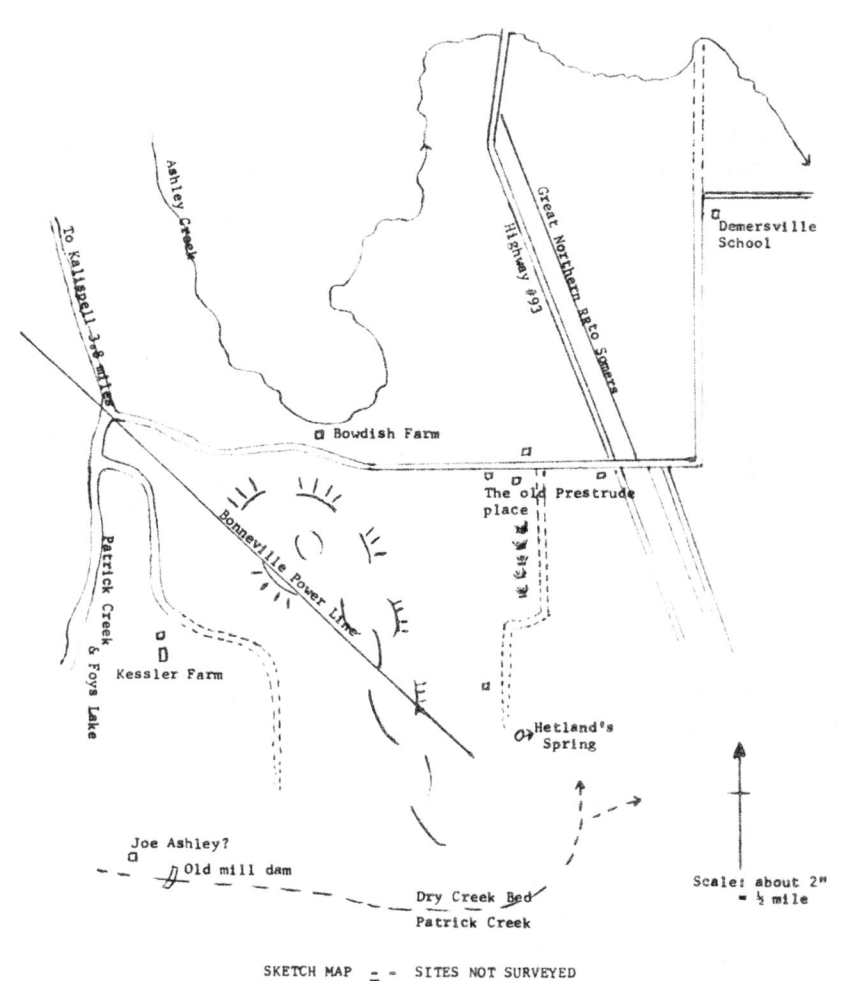

APPENDIX G

BUILDINGS AND METHODS OF CONSTRUCTION

Appendix G is included in order to consider the variety of log cabins and their structure which were used for habitation and storage mainly between the years 1800 and 1900 with considerations towards the rudimentary structure of Howse House.

Had Howse taken it upon himself to erect a winter quarters hewn from native timbers it is not likely the building materials would endure more than 15 to 20 years before rotting off at the ground line.

Several species of trees, if used for substantial construction, would not deteriorate as readily as others providing they were peeled and held in some manner above the ground line. The pitch of the pine and lower portions of the trunks of tamaracks survive the elements rather well. Had Dowse chosen larch for a building material and covered the structure with soil, the timbers would be subject to more rapid deterioration.

During the past three years many sites in the Flathead valley have been explored and surveyed by the authors, However, none of the sites visited revealed any physical evidence of a fireplace, chimney, logs, posts or piles of earth, Rather, most sites were indentations in the earth revealing little surface evidence as to their previous earthly status.

As previously recorded in this manuscript (Historical Part II); Howse House could hardly have been a building of rigorous stature, for it is doubtful this structure was occupied any longer than three months during the latter months of 1810 Perhaps it was only a teepee, a leather tent, or an unglamorous storage cover.

Howse House may have been similar in structure to the buildings at Saleesh House or those at Kullyspell House, but not necessarily comparable in size, Thompson, in his journals, discusses the construction of Saleesh House and Kullyspell House, and from the description given, it is apparent these buildings had posts set into the ground to establish the four corner, and the house brought to completion by shoring-up the walls with 'needles' or props to secure the timbers.

Another laborious method of construction in this era employed the 'toughing' of the corner poles (notching the entire length) and the walls put in place by placing the timbers horizontally between the two upright corner poles.

Despite the record of cabin walls being fabricated by means of placing logs in a vertical position for a stockade enclosure, such type of construction was seldom used.

In the logical process of preparing this Appendix many photographs, sketches. Plans of log cabins, houses, etc., have been consulted, yet none of these items reveal a satisfactory explanation as to the type of structure known as Howse House.

APPENDIX H

INFORMATION FROM LOCAL RESIDENTS

The quest for information on a local basis included interviews with the old timers, their sons and daughters, a search of local newspapers, but little if any knowledge could be determined from memory as many people did not recall nor had they heard of an early day trading post in the Flathead valley. Most information obtained was contemporary; viz. regarding the early day homesteaders.

The Columbia River Historical Expedition of 1921 which was sponsored by the Great Northern Railway in cooperation with the historical societies of several states made the only effort to our knowledge to possibly establish the location of Howse House in the Flathead valley.

On July 24, 1926, an exploratory journey was made by historians, Dr. J. B. Tyrell, and T. C. Elliott; Ralph Budd of the Great Northern Railway, Sidney M. Logen and Sam Johns. This entourage apparently looked for the site of Hone House, but for more intensive purposes sought the point of land where David Thompson first saw Flathead Lake on March 1, 1812. Tyrrell and Elliott established the promontory from which Thompson looked down upon Flathead Lake by data garnered from the Thompson journals. This point is just south of the present town of Polson, Montana.

The same evening Elliot addressed an assemblage at the Elk's temple in Kalispell and cited not only their findings for the day, but a detailed description of Northwestern Montana history. In his address Elliott discussed the Arrowsmith map of 1814:

> "The map of 1814 Arrowsmith as the first to show Flathead Lake and a place named Howse House, just north of the lake where we have been looking today.1 Mr. Tyrrell, with the help of Thompson's map, determined that this may have been at the mouth of Ashley Creek. Now that map of the winter of 1814 ought to be accurate record that Howse as here and traded during the winter of 1810 and 1811, but. We have no copy of House's journal. It may be possible that we find that journal and know what he did over here, but the record of that map is that House, a white man and intelligent, saw Flathead Lake in 1810."

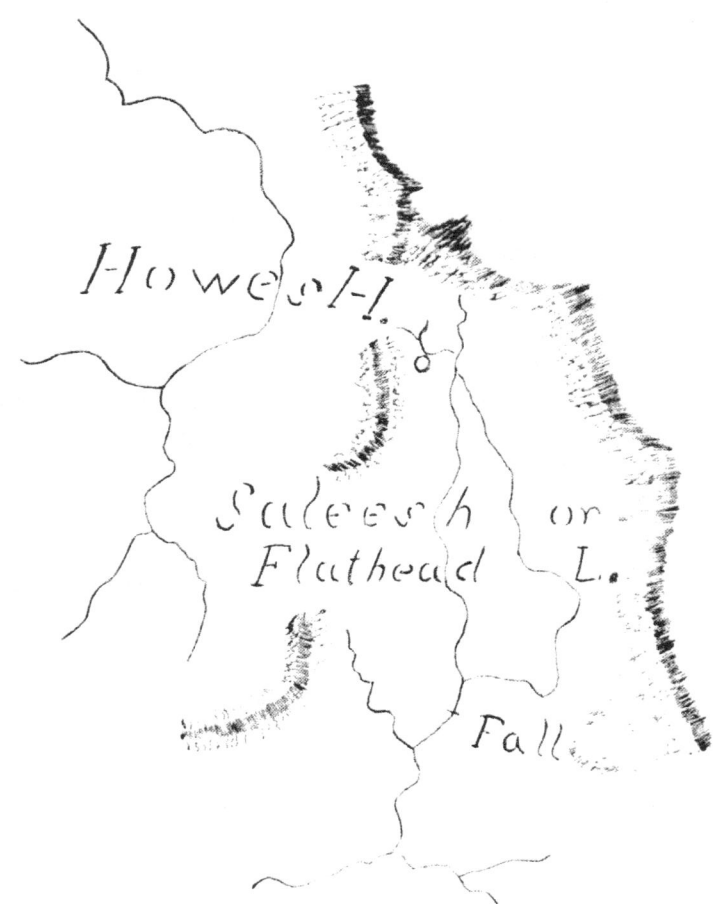

Sketch is taken form Arrowsmith map #6 HII/1000 (1795 - 1819) reproduced and enlarged at a ratio of 1/10 with the use of a Bausch & Lomb cartographers magnifier with millimeter scale. The circle indicating the placement of Howse House is in keeping with the Arrowsmith designation; utilizing a circle for Hudson Bay posts, and two intersecting lines (a cross) for the position of the Northwest Co. posts.

To supplement the information obtained from the Flathead portion of the Columbia River Historical Expedition, Ernest White as interviewed in Sept. of 1959:

> *"The Lockhart farm, where the old homestead is out in the middle of the field, is where I understand Sid Logan and perhaps a man by the name of Tyrrell; perhaps some others came to look for Howe House. This, I believe, was in 1926. I have locked in this field several times & nearly every time I have found some Indian artifacts."*

Lennox Edge, when interviewed in January of 1960, told of guiding a group of interested citizens to the mouth of Ashley Creek; however, he as uncertain whether this was the July 24, 1926, expedition.

The Columbia River Historical Expedition. Presumably sought the general location of Howse House with a brief excursion into the lower Flathead valley. From the statements of Ernest White and Lennox Edge the area of search must have been somewhere between the Lockhart farm and mouth of Ashley Creek. The distance between these two points is about 2 miles. A surface survey as not conducted (by the authors) between these two points; only to follow portions of the old wagon road that connected these two areas.

The S. E. Johns Collection in the Kalispell Public Library Contains several interesting recollections with regards to the possible existence of an early day trading post in the Flathead valley, Mrs. Emma H. Ingalls, a prolific contributor, states:

> *"Joe Ashley, from whom the creek and town take their name, came to the valley in 1857 and took up a homestead where there were the ruins of an old trading post, supposedly built by the Hudson Bay Co. Later he abandoned it and Eugene McCarthy Sr. homesteaded there. Eugene Jr. tells of tearing down the logs and chimney and filling*

> *in the excavation. It is now the David R. Griffith home."*

Tyson D. Duncan, whose numerous recollections are documented throughout the volumes of the Johns collection states:

> *"Joe Ashley had located (on the old McCarthy place) in 1857 but could see nothing in the valley for him and sold his claim for $10. He never returned."*

Again, from the pen of Emma Ingalls:

> *"Ashley Creek was named for Joe Ashley who settled on the old Eugene McCarthy place south of Kalispell in 1857. He left there for the reservation in 1883 and Eugene McCarthy, who had come in as a laborer on the N. P. Ry., then took the place. The remains of the old trading post established in about 1808 were still in evidence. Eugene McCarthy (Police Magistrate) with the scrapers and mules his father had salvaged form the construction work on the N. P. Ry. tore down the old chimney and filled in the excavations, and where this historic building once stood Mr. Griffith has a grain field."*

In 1932, Mrs. Ingalls together with Mrs. E. E. and Day compiled their vivid memoirs in a pamphlet entitled: *Extracts from History of The Flathead Valley*, Dec, 31, 1932. With reference to the Ashley - McCarthy land:

> *"About 1811, David Thompson, of the N/W Trading Co. came into the Flathead and built on what is known as the McCarthy place, south of*

> *Kalispell, a trading post, the ruins of which was still to be seen in 1882. It did not prove successful and was soon abandoned, In 1857, Joe Ashley came into the valley and settled on this land and used these buildings."*

Duncan McCarthy of Kalispell (son of Eugene McCarthy), was interviewed on February 13, 1961. McCarthy contended that the site his father tore down was, to the best of his knowledge, the remains of the store established by Andrew Swaney and Leo Walkup.

The following statement by James K. Lang in the Johns Collection was also reflected by Sidney Logan in his article, *David Thompson and Flathead Lake.*

> *"The Hudson Bay Co, established a trading post in the Flathead valley along about 1844 and placed as manager, Mr. Angus McDonald. This post was located in the lower west side valley near the Bracken school house is now standing, and was maintained for several years after which it was transferred to Post Creek on the Flathead Indians Reservation and conducted by Mr. McDonald for many years. This change was made necessary by the fact that much of the trade with the Indians was being lost as they passed over the mountains to eastern Montana trading posts via the Missoula River route."*

Pursuant to factual and authentic information on the cabin, the following letter by R. J. Ball to T. D. Duncan is more specific and informative:

> *Mr. T. D. Duncan*
> *My dear Friend:*

Your welcome letter about the log cabin received. The log cabin you wish information about, was a short way south of my preemption. It was on what became D. J. Lambert's homestead. When the McCarthy family moved here in 1883, then took the cabin and lived there until McCarthy located and built his home at the point of the mountain, where D. Griffith now lives. I was told the cabin was the remains of an old Indian trading post. You know the old Indian trail used to go around the foot of the mountain by where the cabin stood and kept on until it came to the prairie; around were Andrew Sweeney's store, then on to where Ashley Creek bridge is now, was the ford across Ashley Creek, it was impossible to cross over the swamp below Lambeth house until someone cut the willows out and made a trail across there.

Gene McCarthy, Andrew Swaney and John Foy should be able to give you valuable information about the cabin. It must have stood there a good many years, and one of the first built in the Northwest. It is a pity such old timers as Savaho Ramond Bros., Karl Fisher, John Dooley and others should have passed on and not left valuable information.

That old fellow who ran the ferry at Poison when we first carne here, how much history he could have given. I remember one time his telling me with Fremont in the early forties from St. Louis west, to the Salt Lake, then leaving the party and traveling North until he came to the Flathead. It would be nice, I think, for you to mention our first fourth of July gathering at Foy Lake, 1884. There were just a few of us at that time, and how few of that bunch are here now.

Wishing you success
Yours sincerely
Robert J. Ball

Lester Foy told of arriving in the Flathead valley together with his parents (Mr. & Mrs. John M. Foy) and two brothers and five sisters on September 10th, 1883 and spending the winter at a point somewhere between the present road which leads from Ball's crossing to the foothill road and Ashley Creek. Foy added that the Eugene McCarthy family arrived in October of the same year (1883). Foy stated:

> *"I can remember the Indians who were on their way north, and settled just south of us. I remember them pounding their tom-toms all night and all day; taking turns, working for a chinook."*

Foy said he was nine years old at the time, but did not remember a cabin in the vicinity of which Ball mentions above, but recalled that it took his father and McCarthy a total of three weeks to make a round trip to Missoula for supplies.

The following is a portion of a letter addressed to W. H. Lawrence, Supt. of the Water Department, Kalispell, Montana:

> *"Now as all fur traders, fur buyers and fur companies came from the north, this being good fur and buffalo country, they thought the boundary* [49th parallel] *was much farther south of where it is now. I know, for I have found at Lary & Lynch ranch, near the mouth of Flathead*

River, what proved to be the ruins of an old Hudson's Bay Trading Post. Mr., Lary came here over 50 years ago and he told me then that the old boundary line was the north end of Flathead Lake."

HENRY BOSE

Considering the repeated but reliable testimony offered by the early settlers, some credence must be given to the statements relating to an early day trading post on the Ashley McCarthy - Griffith, or the Lambert property.

A discerning observation among archaeologists would conveniently rule in this instance; *'that settlement can be predicted where previous habitation had existed.'*

According to the record, Ashley settled in the alleged trading post, and was followed in turn by McCarthy who also utilized the building as a pro tem residence. The fact that both of these parties found this area agreeable, and accessible, is not to be overlooked.

Further, the early roads and/or trails in this particular area followed more or less the high ground to avoid the circuitous route imposed by the numerous and impassable swamps on the valley floor, The spacious and accommodating terrace provided by the slope of the foothills in this area, not only afforded ready access to wood, but an abundant water supply from the present Hetland springs, Patrick Creek, and nearby Ashley Creek.

All of this would tend to confirm that this area was the most favorable for existence under the most primitive conditions.

Made in the USA
Monee, IL
12 February 2024